T0288930

The Role of Southeast Asia in U.S. Strategy Toward China

Richard Sokolsky, Angel Rabasa, C.R. Neu

Prepared for the United States Air Force

RAND

China's geopolitical ambitions will play a crucial role in shaping the future of Southeast Asia and the U.S. military posture in the region. There are a number of strategic directions China could take depending on which domestic and external factors emerge as key determinants of Chinese national security policy. Which path China will follow remains unknown, however, and this uncertainty complicates the formulation of an effective policy for managing China's rising power throughout the Asia-Pacific region.

Coping with this uncertainty poses a particularly difficult dilemma for U.S. security strategy in Southeast Asia. Many Southeast Asian states are concerned about the growth of Chinese military capabilities and China's long-term intentions. At the same time, these countries have a healthy fear of provoking China and lingering doubts about the credibility of U.S. security commitments. These ambivalent attitudes and threat perceptions, combined with intra-regional tensions, present both opportunities and challenges for expanded U.S. military cooperation with the Association of Southeast Asian Nations (ASEAN) states.

This study examines the role of the ASEAN countries in U.S. security strategy toward China. It focuses in particular on regional perceptions of a "rising China" and the possibilities for enhanced U.S. military cooperation with the countries of Southeast Asia in an uncertain and potentially unstable environment.

This research was conducted in the Strategy and Doctrine Program of Project AIR FORCE under the sponsorship of the Deputy Chief of Staff for Air and Space Operations, U.S. Air Force, and the Commander, Pacific Air Forces. The report should be of value to the national security community and to those in the general public who are interested in United States-Chinese relations and the future of the Asia-Pacific region. Comments are welcome and should be sent to the authors or the project leader, Dr. Zalmay Khalilzad.

PROJECT AIR FORCE

Project AIR FORCE, a division of RAND, is the United States Air Force's Federally Funded Research and Development Center (FFRDC) for studies and analyses. It provides the Air Force with independent analyses of policy alternatives affecting the development, employment, combat readiness, and support of current and future aerospace forces. Research is performed in four programs: Aerospace Force Development; Manpower, Personnel, and Training; Resource Management; and Strategy and Doctrine.

CONTENTS

FIGURES

TABLES

SUMMARY

Chinese behavior in Southeast Asia and the growth of Chinese military capabilities have aroused apprehension among countries of the Association of Southeast Asian Nations (ASEAN). How they react to a "rising China" could have major implications for U.S. interests, military requirements, and the USAF posture in the region.

This study examines the implications of a rising China for U.S. security strategy and defense planning in Southeast Asia. In particular, it addresses the following questions: What role are the ASEAN states likely to play in developing a hedge against the possible emergence of an overtly aggressive China? If China emerges as a hostile competitor, are the ASEAN states likely to contribute to a United States-led effort to deter or oppose a Chinese challenge to regional security? What is the most effective strategy for pursuing cooperative military arrangements with the ASEAN states?

COPING WITH UNCERTAINTY

A host of internal and external factors will influence China's behavior over the coming years, and it is difficult to predict which will be the primary determinants of Chinese foreign policy. Several alternative paths are open to China, ranging from aggressive nationalism and hostility to U.S. interests in the Asia-Pacific region and beyond to pragmatism and partnership with the United States.[1]

---

[1]For a discussion of the determinants of Chinese national security behavior, see Z. Khalilzad et al., *The United States and a Rising China: Strategic and Military Implications*, RAND, MR-1082-AF, 1999.

China currently has a strong stake in maintaining both good relations with its neighbors and the United States and a stable environment in the Asia-Pacific region. Nevertheless, China's intentions and
ambitions could change over the longer term, particularly if China
assumes its place among the world's leading economic and technological powers. Under these circumstances, other determinants of
Chinese behavior, including the desire for regional hegemony and
"national redemption," could lead a more powerful and hostile
China to mount an aggressive challenge to the United States for
global and regional primacy.

Much of the anxiety about a rising China and the potential Chinese
challenge to American dominance of the Asia-Pacific region stems
from Chinese behavior in Southeast Asia—in particular, its use of
force to defend Chinese territorial claims and continued Chinese development of power projection capabilities. At the same time, the
future direction of Chinese policy toward the region remains uncertain. Although territorial conflicts in the South China Sea could lead
to armed conflict, the Chinese have strong incentives to resolve these
disputes without the use of force. Accordingly, the Chinese have engaged in a pattern of "creeping irredentism," steadily pressing their
claims while avoiding actions that might provoke a large-scale
military engagement and the formation of an anti-Chinese bloc. A
particular challenge for the United States will be to devise an
effective strategy for responding to more limited and ambiguous
Chinese military challenges.

"CONGAGEMENT" IN SOUTHEAST ASIA

Uncertainty over China's future geopolitical orientation complicates
U.S. policy toward China and American defense planning in East
Asia. On the one hand, the current U.S. policy of engagement seeks
to take advantage of opportunities for cooperation with China but
could leave the United States unprepared to deal with the emergence
of a hostile China. On the other hand, a policy of containment, as
some observers have advocated, might better prepare the United
States to deal with the emergence of an adversarial China. But it may
also squander the benefits of cooperation and, more important, by
treating China as an enemy precipitate the very outcome the United
States wishes to avoid.

Southeast Asia is likely to prove a critical testing ground for implementing a "third way" of dealing with China's rising power—what might be called a strategy of "congagement" that seeks to integrate China into the international system while both deterring and preparing for a possible Chinese challenge to it.[2] Given the uncertainties about China's future strategic orientation and the divergent views of China within ASEAN, implementing the hedging part of the "congagement" strategy will present difficulties for the United States and the ASEAN countries. The key issue is not whether the United States should seek to establish a prudent hedge in Southeast Asia against the possibility of an adversarial China, but rather how to manage the implementation of a hedge strategy—the priority, content, and timing of hedging actions, the appropriate balance between the engagement and hedging elements of U.S. strategy, and the resources that should be expended to establish a hedge.

IMPLICATIONS FOR U.S. DEFENSE PLANNING AND THE USAF

Applying the principles of the "third way" policy would have the following implications for the United States and the USAF:

- **Regional basing and access.** The United States faces political and resource constraints in creating a regional infrastructure to support large-scale conventional military operations. Nonetheless, it may be possible to secure cooperation from several ASEAN countries in establishing a more robust network of access arrangements. The Philippines and Singapore are the most promising candidates for such enhanced access.

- **Military operations and force structure.** The USAF should consider the merits of increasing exercises in and rotational deployments of combat aircraft to Southeast Asia. It might also be worthwhile to begin a quiet dialogue with friendly countries on joint cooperation in meeting USAF operational requirements for responding to contingencies of common concern in the region. Such a dialogue could include discussions of how U.S. arms transfers and combined exercises could promote interoperability with ASEAN forces.

---

[2]For a more detailed description of this strategy, see Khalilzad et al.

- **Shaping activities.** For at least the next 5–10 years, the United States will have an opportunity to cultivate stronger military ties with many ASEAN states. Military-to-military contacts should emphasize encouraging professionalism and modernization in a democratic context. This will be particularly important in Indonesia, where military-to-military contacts could help to "acculturate" the Indonesian armed forces (TNI) in democratic civilian control of the military and assist in the transition to a military doctrine consistent with Indonesia's democratic evolution. Activities could include technical assistance for doctrine development, joint force operations, logistics and maintenance support, and training to combat smuggling, piracy, and drug trafficking. Regional states could be encouraged to develop an integrated air defense network.

- **A tailored strategy.** The United States should adopt an incremental approach to hedging. The initial phase of a hedging strategy should focus on peacetime military engagement, dialogue, reassurance, and trust-building. Given the constraints and uncertainties associated with expanded U.S. access to facilities in ASEAN countries, the United States/USAF should adopt a "portfolio" approach—in other words, the United States should diversify its regional military infrastructure as much as possible to hedge against loss of access in any one country and seek to strengthen military ties with the Philippines, Singapore, Thailand, Malaysia, Indonesia, and Vietnam.

The level of democracy and human rights practices in the respective ASEAN countries will pose potential political constraints on a U.S. engagement strategy with the ASEAN militaries. The military's involvement in political and internal security activities in some ASEAN countries, particularly Indonesia, has created substantial barriers in the past to military-to-military cooperation with the United States. At the same time, the militaries in most ASEAN countries are important, and sometimes dominant, players in the political system, as well as in defense and security policy decisions. The United States therefore needs to walk a fine line between engaging ASEAN militaries in order to influence their values, security doctrines, and political actions and thereby advance U.S. strategic interests in the region and avoiding association with activities incompatible with U.S. values. At the same time, there should be a realistic understanding of the limits of U.S. influence and its ability to dictate outcomes on issues of paramount importance to regional governments and militaries.

ACKNOWLEDGMENTS

The authors wish to thank all those who contributed directly or indirectly to this study. Within RAND, we thank Dr. Zalmay Khalilzad, Corporate Chair in International Security and Director of Strategy and Doctrine, Project AIR FORCE. Dr. Khalilzad supervised a series of studies on Asian security issues that provided the intellectual framework for this project, and participated in discussions with government officials and defense experts in Southeast Asia that became an important part of our data base. We also thank our research assistant, Andrew Mok, who contributed to the chapter on U.S. interests; Shirley Lithgow, who patiently prepared several early versions of this report; and Joanna Alberdeston for her work on administrative details and coordination of the manuscript. In addition, the authors appreciate the invaluable contribution of our editor, Jeanne Heller, and our production editor, Alisha Pitts, in turning the manuscript into a finished product.

We are particularly grateful to our distinguished outside reviewers: Dr. Karl Jackson, Director of the Southeast Asia Studies Program at the Paul H. Nitze School of Advanced International Studies, Johns Hopkins University; and Dr. James Clad, Research Professor of Southeast Asian Studies at the School of Foreign Service, Georgetown University. The study benefited immensely from the perceptive comments and insights provided by Dr. Jackson and Dr. Clad. Any shortcomings are, of course, entirely our responsibility. Our thanks extend to government officials and academics in Southeast Asia for their discussions with the authors during visits to Southeast Asia. We thank in particular the staff of the National Defense College of the Philippines, the Institute of Defence and Strategic Studies of Singapore, the National Defense College of Thailand, and the

National Resilience Institute (LEMHANNAS) and the Centre for Strategic and International Studies of Indonesia.

Chapter One

INTRODUCTION

China's emergence as a major regional power over the next 10 to 15 years could intensify United States-People's Republic of China (PRC) competition in Southeast Asia and increase the potential for armed conflict. The United States is currently the dominant extraregional power in Southeast Asia. The Association of Southeast Asian Nations (ASEAN) continues to rely on U.S. military forces to guarantee regional stability and security and to balance China's growing power. Economic growth in the Asia-Pacific region, which is important to the economic security and well-being of the United States and other powers, depends on preserving American presence and influence in the region and unrestricted access to sea-lanes.

At the same time, China seeks to reassert its historical role as the dominant regional power and has substantial irredentist territorial and maritime claims in the South China Sea. Over the next decades, China's dependence on imported oil will increase substantially, and much of this oil will transit Southeast Asian sea-lanes.[1] The Chinese are likely, therefore, to pursue an activist policy to influence regional developments as well as to acquire the capabilities to project military force throughout the region. In short, the United States and China have very different concepts of how security should be organized in the Asia-Pacific region and these competing visions could clash in Southeast Asia.

---

[1]The China National Petroleum Corporation (CNPC) and the China National Offshore Oil Corporation (CNOOC) hold minority shares of offshore production and exploration blocks in the Strait of Malacca and the Java Sea.

It is by no means axiomatic, however, that a "rising China" will pose a military threat to the ASEAN states or lead inexorably to military conflict with the United States or a Chinese geopolitical challenge to American primacy in the Western Pacific. Chinese policy in Southeast Asia is the result of a complex and shifting mosaic of conflicting factors, including China's internal political and economic evolution, regional security dynamics, the growing economic interdependence between the ASEAN states and China, and the U.S. military posture and security policy in the region. The real issue is the extent to which regional countries would feel the need to accommodate China's ascendancy—military threat is only one element in the panoply of Chinese policy instruments.[2]

China has territorial, maritime, and security goals that would extend its presence into the heart of Southeast Asia. In particular, China claims sovereignty over the entire South China Sea as well as the Spratly and Paracel islands, which sit astride vital sea-lanes through which 25 percent of the world's shipping passes. China's claims to the South China Sea fall into the same category as its claims to Taiwan and Tibet—as a sovereign part of China—and exceed in scope (surface, subsurface, fisheries, air rights) those of other claimants.[3]

The Chinese have aggressively defended these claims with a tough declaratory policy backed with the use of force. Moreover, China is steadily improving its ability to project force throughout the South China Sea. Absent profound changes in China's political system or the creation of an effective regional security "architecture," there are relatively few political, legal, or institutional constraints on China's use of force or coercion to pursue its interests.[4] But China also seeks to create a stable environment in Southeast Asia for the expansion of trade and investment, which undergirds China's strategy of market-

---

[2]For an illuminating discussion of the bifurcated nature of Chinese policy in Southeast Asia, see Wayne Bert, "Chinese Policies and U.S. Interests in Southeast Asia," *Asian Survey*, Vol. 33, No. 3, March 1993.

[3]Dr. James Clad's comments to authors, January 2000.

[4]Jonathan D. Pollack, "Designing a New American Security Strategy for Asia," in James Shinn (ed.), *Weaving the Net: Conditional Engagement with China*, New York: Council on Foreign Relations, 1996, p. 118.

led economic growth and, with the demise of communist ideology, the country's political stability and Chinese Communist party rule.

It is possible, therefore, but not inevitable that over the next 15 years China could emerge as a more powerful and hostile competitor to the United States for regional influence and access. Deterring armed conflict between China and ASEAN states and a Chinese challenge to the existing security order in Southeast Asia will pose a major security challenge to the United States. In light of the weakness of individual ASEAN countries and the limitations on effective regional defense and security cooperation, for the foreseeable future only the United States will be capable of preventing China from achieving regional hegemony should Beijing move in this direction. As a consequence, it is plausible that some of the ASEAN states, if confronted by the prospects of an aggressive and threatening China, may rely increasingly upon the United States and its military power for deterrence, reassurance, and protection.[5]

This study examines the role of Southeast Asia in a U.S. strategy toward China predicated on the notion that, given the many paths China could take, the most prudent approach to managing a rising China is to engage but hedge against an uncertain future. Chapter Two provides an overview of U.S. strategic interests and objectives in Southeast Asia. Subsequent chapters examine the nature of potential Chinese military threats to U.S. interests in the region, the attitudes of the ASEAN states toward military cooperation with each other and the United States, the likely response of the ASEAN states to a rising China, and the prospects that regional security arrangements might help prevent conflict and constrain China. The concluding chapter addresses the implications for U.S. defense planning, military presence, and force posture in Southeast Asia.

---

[5]The view that ASEAN's expansion may have weakened ASEAN's leverage vis-à-vis China by diluting the organization's cohesion and reducing the chances for consensus has gained wide acceptance among security experts. With the admission of Cambodia on April 30, 1999, ASEAN now consists of the following ten nations: Brunei, Cambodia, Indonesia, Laos, Malaysia, Burma (Myanmar), the Philippines, Singapore, Thailand, and Vietnam.

Chapter Two

U.S. OBJECTIVES AND INTERESTS IN SOUTHEAST ASIA

The United States has strong economic and strategic stakes in Southeast Asia. Indeed, ASEAN has eclipsed the importance of several traditional U.S. trading partners. Moreover, notwithstanding the end of the Cold War, U.S. influence and credibility in the region and beyond continue to depend on America's ability to honor its security commitments and defend the principle of freedom of navigation.

ECONOMIC STAKES

ASEAN, with a population of over 500 million, is a large market for American goods and services as well as an increasingly important U.S. investment destination and source of imports—for example, ASEAN exports to the United States in 1998 totaled about $60 billion. The region's geographic location astride sea-lanes connecting not only the Indian and Pacific Oceans but also north-south routes linking Australia and New Zealand to the countries of Northeast Asia also imbues this region with strategic relevance for international security and commerce. If national security interests are more broadly defined to include exposure to drug trafficking and international crime, the role of Southeast Asia as a leading source of drugs makes this region one that demands sustained attention. Finally, as those of Southeast Asian origin in the United States increase, so too do cultural and social ties grow between the United States and Southeast Asia.

Despite the turbulence caused by the Asian economic crisis, ASEAN remains an important U.S. trading partner—the destination of some $40 billion in U.S. merchandise exports and the source of $78 billion in U.S. merchandise imports in 1999.[1] In recent years, ASEAN has been second only to Japan and well ahead of China, Hong Kong, and Korea in terms of U.S. merchandise exports to the Pacific Rim. The level of U.S. merchandise imports from ASEAN has been less than from Japan, but similar to the level of imports from China, and much greater than imports from Korea and Hong Kong (see Table 2.1).

U.S. exports to ASEAN are significant and expected to resume robust growth when regional purchasing power recovers from the effects of the financial crisis.[2] There is potential for further growth in United States-ASEAN trade as structural changes in the region's economies increase the value of the service and high-tech sectors, where U.S.

Table 2.1

U.S. Trade with Asia: 1996–1999
(millions of U.S. dollars)

Year	China	Japan	Hong Kong	Korea	ASEAN
			Exports		
1996	11.933	67.607	13.966	26.621	43.631
1997	12.862	65.549	15.117	25.046	48.271
1998	14.241	57.831	12.925	16.486	39.37
1999	13.118	57.484	12.647	22.954	39.873
Sum					171.145
			Imports		
1996	51.513	115.187	9.865	22.655	66.427
1997	62.558	121.663	10.288	23.173	71.014
1998	71.169	121.845	10.538	23.942	73.395
1999	81.786	131.404	10.531	31.262	77.67
Sum					288.506

SOURCE: U.S. Department of Commerce, Bureau of the Census, Foreign Trade Division, www.census.gov/foreign-trade/balance/index.html

[1]U.S. Department of Commerce, Bureau of the Census, Foreign Trade Division, at www.census.gov/foreign-trade/balance/index.html

[2]U.S.-ASEAN Business Council, *ASEAN Market Overview*, June 1999, p. 2.

companies have competitive advantages. Furthermore, although in absolute terms ASEAN's total nominal gross domestic product (GDP) in 1997 of about $675 billion is roughly comparable to that of one of the largest American states,[3] this figure belies the importance and attractiveness of ASEAN as a market for American goods and services. This total GDP figure, and related per capita income statistic of about $1300, masks the exceptionally wide range of per capita incomes, ranging from a low of $250 in Vietnam to a high of more than $20,000 in Singapore and Brunei,[4] as well as diverse levels of socioeconomic development. Because these markets range from the largely agricultural to the highly sophisticated post-industrial, they complement the diversity of goods and services produced by the U.S. economy and therefore offer a well-balanced growth opportunity for American exports.

ASEAN is also a major destination for American foreign investment, as measured by foreign direct-investment positions. From 1990 to 1997, U.S. foreign direct investment in the region climbed from $11.8 billion to $37.5 billion, surpassing U.S. investment in both Japan and Brazil of $35.6 billion and $35.7 billion, respectively.[5] Perhaps more surprising, U.S. direct investment in ASEAN was seven times greater than that in China and almost double that in Hong Kong.[6] While the Asian financial crisis sharply reduced capital inflows to the ASEAN countries, most observers expect the return of portfolio capital to regional equity markets as the recovery takes hold.[7]

The changing complexion of American investment in ASEAN also points to the growing economic importance of this region to the United States. In the 1980s, most investment centered on the oil and

---

[3]According to the U.S. Department of Commerce, *Statistical Abstract of the United States*, 1998, No. 719, the gross state product of California and New York in 1996 were $880 billion and $563 billion, respectively.

[4]U.S.-ASEAN Business Council, p. 2.

[5]*Statistical Abstract*, 1998, No. 1312; *Survey of Current Business*, July 1998, p. 43, Table 3.2. Estimates of portfolio investment are problematic. The fact that U.S. direct investments in ASEAN surpass those in Japan may be partially explained by Japanese restrictions on large equity investments by foreigners, which makes it harder for U.S. investors to meet the U.S. definition of direct investment (10 percent ownership).

[6]*Statistical Abstract*, 1998, No. 1312.

[7]See Asian Development Bank, *Asian Development Outlook 1999*, Manila, p. 7.

gas sector. However, in 1997, oil and gas had fallen to 29 percent of total investment while manufacturing and services accounted for 37 percent and 34 percent, respectively.[8] Sustained growth in per capita incomes as well as the trade and investment liberalization measures in the decade before the onset of the economic crisis not only increased the number of American companies with operations in ASEAN but also encouraged existing investors to increase their exposure to Southeast Asia.[9] Furthermore, by drying up many domestic sources of capital and bringing down asset prices, the Asian financial crisis opened new opportunities for American investors who were previously excluded from the region.

Economic restructuring, however, has been uneven across the ASEAN landscape. In some countries, the economic crisis prompted economic reforms that may lay the basis for sustained economic growth. For instance, the ASEAN member countries ratified two agreements under the ASEAN Framework Agreement on Services that call for liberalization of regional trade in sectors such as air transport, telecommunications, and financial services, with the ultimate intention of providing improved market access.[10] On the other hand, weak bankruptcy laws in several countries made it difficult for creditors to force insolvent companies into liquidation. As a result, regional economies have not been able to benefit fully from the inflow of new capital seeking investment opportunities or from the resumption of bank lending. The Indochinese countries and to some extent Malaysia have been unwilling to undertake major economic reforms and have coasted on the growth of the global economy. There has been significant restructuring of the banking sector in Thailand but very little in Malaysia, and in Indonesia the banking and corporate sectors are in complete disarray. The risk to the

---

[8]U.S.-ASEAN Business Council, p. 5.

[9]One indicator of the range of commercial enterprises with interest in Southeast Asia is the membership roster of the U.S.-ASEAN Business Council, which includes companies like Aetna International, Bank of America, Bell Atlantic, Keystone Foods, Motorola, and Pfizer International.

[10]U.S.-ASEAN Business Council, p. 11.

ASEAN economies is that, absent structural economic reform, the current recovery may not be sustainable.[11]

A final economic consideration has important implications for political relationships in Southeast Asia. Over the period 1995–1998 (the last four years for which data are available), the United States accounted for 19 percent of merchandise exports from ASEAN countries. Japan accounted for a further 14 percent. China and Hong Kong accounted for only 3 and 6 percent, respectively, with Korea accounting for another 3 percent (see Figure 2.1). The relative positions of the United States and Japan are reversed as sources of merchandise imports by the ASEAN countries. During the same

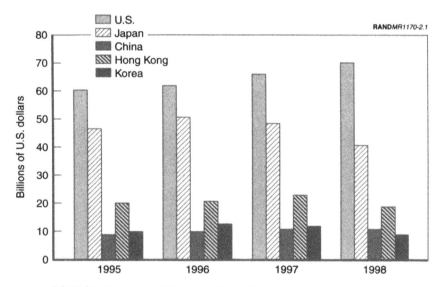

SOURCE: International Monetary Fund, *Direction of Trade Yearbook*, 1999.

Figure 2.1—ASEAN Merchandise Exports

_____

[11]The resistance to reform by entrenched political and business institutions is illustrated in Mark L. Clifford and Pete Engardio, *Meltdown: Asia's Boom, Bust, and Beyond*, New Jersey: Prentice Hall, 2000.

four years, 15 percent of ASEAN merchandise imports came from the United States, 21 percent from Japan, 4 percent from the PRC, 3 percent from Hong Kong, and 5 percent from Korea (see Figure 2.2). Assuming no dramatic reordering of this trade structure, the United States and Japan are by far the most important international trade partners for the ASEAN nations. While such trade relationships do not necessarily determine policy outcomes, extensive trade and investment ties can contribute to a coincidence of strategic interests.

SEA-LANES

Southeast Asia lies at the intersection of two of the world's most heavily traveled sea-lanes. The east-west route connects the Indian and Pacific Oceans, while the north-south one links Australia and New Zealand to Northeast Asia. Both routes are economic lifelines by which the economies of Northeast Asia receive critical inputs like oil and other natural resources and export finished goods to the

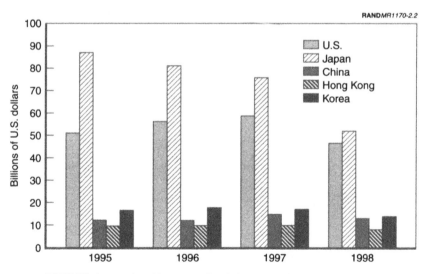

SOURCE: International Monetary Fund, *Direction of Trade Yearbook*, 1999.

Figure 2.2—ASEAN Merchandise Imports

rest of the world. Also, because of the littoral nature of most ASEAN nations and the relative lack of land-based transport, much intraregional trade also depends on these waterways. From a military perspective, these sea-lanes are critical to the movement of U.S. forces from the Western Pacific to the Indian Ocean and the Persian Gulf. Furthermore, almost all shipping must pass through one of three straits, or "chokepoints," in the region: the Strait of Malacca, Sunda Strait, and the Straits of Lombok and Makassar (see Figure 2.3.).[12] Three regional states—Indonesia, Malaysia, and Singapore—sit astride or are adjacent to these chokepoints and thus are able to exercise potential control over a significant percentage of the entire world's maritime trade.[13]

During the Cold War, maintaining freedom of navigation of these waterways for American military vessels while denying that same freedom to the Soviet Union in the event of a conflict was the top American strategic objective; facilitating seaborne commerce was a secondary goal. With the demise of a clear and immediate global military threat, economic considerations have become more salient. Nonetheless, the United States and its regional friends must still pay attention to a range of potential threats, both conventional and non-conventional, to freedom of navigation and the sea-lanes and retain the capability to deny freedom of operation to potential adversaries.

In addition to the ability to counter military threats to freedom of navigation, the United States has a strategic interest in maintaining confidence among all East Asian states that it remains a reliable guarantor for universal freedom of navigation. While the proportion of U.S. trade going through these waterways is small, American allies such as Japan, Korea, Australia, and the Philippines and friendly

---

[12]Although a casual inspection of maps of the region presents a number of potential routes through Southeast Asia, the type of ships (e.g., supertankers) and navigational hazards (e.g., reefs and shallow waters) generally rule out passages other than those described above, plus possibly the straits east of East Timor and between Sulawesi, Buru, and Ternate, through the Banda and Molucca Seas.

[13]For example, John H. Noer notes in *Chokepoints: Maritime Economic Concerns in Southeast Asia*, Washington, D.C.: National Defense University, 1996, p. 3, that shipping traffic through the Malacca Strait is several times greater than through either the Suez or Panama canals.

RANDMR1170-2.3

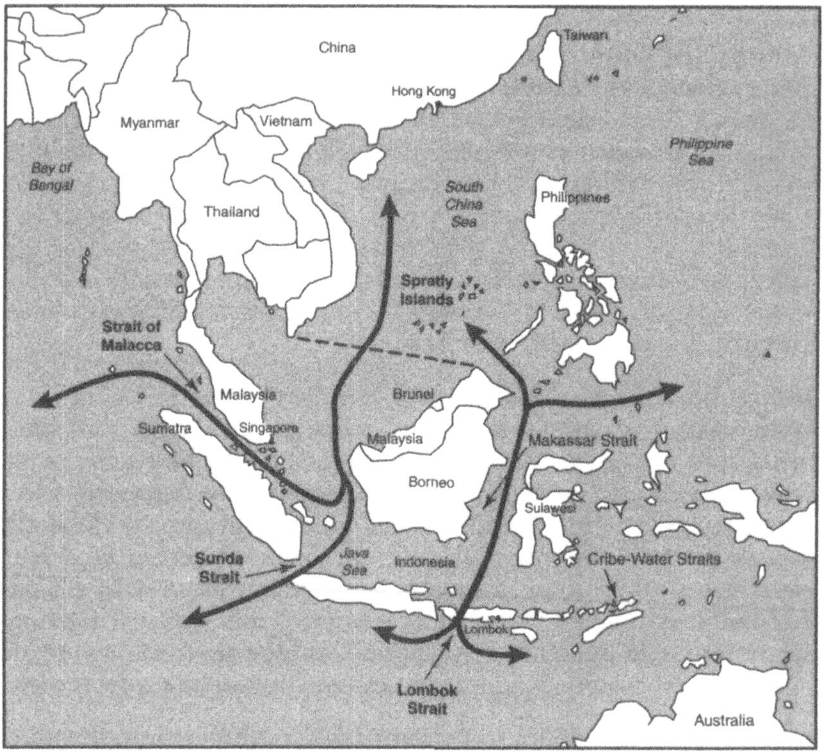

SOURCE: John H. Noer, *Chokepoints: Maritime Economic Concerns in Southeast Asia*, National Defense University, Washington, D.C.

Figure 2.3—Strategic Chokepoints: Straits of Malacca, Sunda, and Lombok, and Sea-Lanes Passing the Spratly Islands

states such as Singapore depend on the Southeast Asian sea-lanes. For example, over 40 percent of trade from Japan, Australia, and ASEAN transits these chokepoints. The comparable figure for Hong Kong, Taiwan, and Korea is over 25 percent (also see Tables 2.2 and 2.3).[14]

_____

[14]Noer, pp. 3–4.

Table 2.2

Maritime Exports in the Southeast Asian Sea-Lanes, 1993[a]

Economy	Tons[b] (millions)	Value ($ billions)	Percentage of Export Value
Japan	33.6	153	42.4
NIEs[c]	24.7	78	25.7
Australia	133.6	17	39.5
China	8.9	20	21.8
Europe[d]	40.8	107	6.8
Southeast Asia	171.2	114	55.4
United States	11.1	15	3.3
World	830.0	568	15.1

SOURCE: Noer, *Chokepoints: Maritime Economic Concerns in Southeast Asia,* 1996.

[a]Interregional cargoes that passed through the Straits of Malacca, Sunda, or Lombok, or by the Spratly Islands.

[b]All tons are metric tons, also called "long tons."

[c]Newly Industrialized Economies: Korea, Taiwan, Hong Kong.

[d]Excludes eastern Europe and Mediterranean regions.

Table 2.3

Maritime Imports in the Southeast Asian Sea-Lanes, 1993[a]

Economy	Tons[b] (millions)	Value ($ billions)	Percentage of Import Value
Japan	385.0	102	42.0
NIEs[c]	199.8	85	28.3
Australia	10.2	24	52.8
China	23.0	11	10.3
Europe[d]	41.7	162	10.5
Southeast Asia	139.4	118	52.5
United States	9.5	27	4.5
World	830.0	568	15.2

SOURCE: Noer, *Chokepoints: Maritime Economic Concerns in Southeast Asia,* 1996.

[a]Interregional cargoes that passed through the Straits of Malacca, Sunda, or Lombok, or by the Spratly Islands.

[b]All tons are metric tons, also called "long tons."

[c]Newly Industrialized Economies: Korea, Taiwan, Hong Kong.

[d]Excludes eastern Europe and Mediterranean regions.

An overt attempt by a hostile power to control or interdict the sea-lanes is not the only threat to freedom of navigation through the straits. A number of unconventional threats also deserve attention. Piracy has long plagued ships traversing the region. In addition, the combination of heavy traffic through the straits and deep-draft vessels (i.e., heavily laden ships that have little maneuverability because deep channels in the straits are usually narrow as well) increases the risks of collisions and attendant environmental damage. Because the Strait of Malacca, the primary channel traversed by supertankers, is considered an international waterway and is governed by the law of the sea, the countries most directly affected by the environmental impact of a collision—Indonesia, Singapore, and Malaysia—have no regulatory authority over safety of navigation on the strait.[15]

Terrorist acts are another potential threat to freedom of navigation. The deliberate sinking of a ship in any of the straits could cause significant disruptions. While it is not clear that the economic cost of a temporary disruption of shipping on the straits would be large, there would be political fallout from any such deliberate attempt at disruption. In that sense, these chokepoints are not entirely physical but figurative. The more important role these waterways play is their symbolic value. Consequently, they present the United States with an opportunity to demonstrate its commitment to the region's security. By helping to ensure freedom of navigation, the United States can provide comfort to regional states and discourage extraregional actors from attempting to exert influence in ways that are detrimental to overall regional security.[16]

---

[15]There is, however, a trilateral strait maritime traffic consultative system between Indonesia, Malaysia, and Singapore.

[16]The United States, of course, is not alone in contributing to freedom of navigation in the straits. The contributions of the navies of the ASEAN countries and the Five Power Defense Agreement countries should also be noted in this regard.

Chapter Three

CHINA'S POTENTIAL MILITARY THREAT TO SOUTHEAST ASIA

The potential threats China poses to Southeast Asia can be placed in two broad categories: conventional military threats and more ambiguous and subtle challenges, possibly in the guise of maintaining regional order. We recognize that these categories are part of a continuum: China employs both approaches as needed—beginning with subtle or indirect threats, and escalating when that approach fails or where a "lesson" is required.[1] Nevertheless, the distinction between the two kinds of threats is analytically useful: threats in the former category are easily identifiable and therefore more amenable to deliberate planning, including deterrence and response. Because of their lower profile and ambiguous nature, threats in the latter category may be more likely to materialize but harder to anticipate or counter effectively.

CONVENTIONAL MILITARY THREATS

There are two conventional military threats that would require a U.S. diplomatic or military response:

* An aggressive and hegemonic China could threaten freedom of navigation in the South China Sea, perhaps to coerce the United States, Japan, or the ASEAN states into accepting Chinese politi-

---

[1]Dr. James Clad's comments to authors, January 2000. For a more extensive analysis, see Mark Burles and Abram N. Shulsky, *Patterns in China's Use of Force*, RAND, MR-1160-AF, 2000.

cal demands. If faced with this prospect, the United States might seek support from individual ASEAN states to carry out a defense of the sea-lanes, or one of the ASEAN states might request such U.S. assistance. While U.S. naval forces would play the primary role in such a contingency, U.S. air power might also be called upon to protect U.S. naval forces or the territories and facilities of the ASEAN states against Chinese military attacks.

*   China could try to forcibly establish and maintain physical control over all or most of the Spratly Islands, prompting requests for military assistance from one or more of the ASEAN countries. Such a Chinese operation could feature the threat or use of force against the territory of an ASEAN state, either to compel acceptance of Chinese demands or to defeat opposing military forces; alternatively, China could expand its "salami tactics" to assert control over more territory. Under either of these circumstances, ASEAN governments could request a more visible and substantial U.S. military presence, including emergency deployments of U.S. naval vessels and combat aircraft as a demonstration of America's commitment to use force to meet its security commitments.

Threats to the Southeast Asian SLOCs

For a host of reasons, the likelihood of an overt Chinese attempt to disrupt Southeast Asian sea-lanes over a sustained period of time would appear to be low. First, under normal circumstances, China has strong economic incentives to maintain freedom of navigation for its own shipborne commerce through Southeast Asian sea-lanes. Over $1 trillion in trade passes through these sea-lanes each year. China's share of this trade, including trade that transits Hong Kong, is close to $100 billion a year—or roughly 16 percent of China's GDP—and growing at an annual rate of over 16 percent.[2] Moreover, China's dependence on these sea-lanes is expected to grow, especially for imported oil: by the year 2015, according to several forecasts, China's demand for energy is projected to increase 160 percent and Chinese consumption of Persian Gulf oil, which would pass

---

[2]Henry J. Kenny, *An Analysis of Possible Threats to Shipping in Key Southeast Asian Sea Lanes*, Alexandria, VA: Center for Naval Analyses, 1996, pp. 38–39.

through Southeast Asian sea-lanes, is expected to triple. Thus, a serious and prolonged blockage of Southeast Asian sea-lanes would inflict damage on Chinese economic growth by cutting off trade to China and that of China's key trading partners in the Asia-Pacific region.

Second, although Chinese naval forces might engage in "police actions" to combat piracy, political considerations would discourage Chinese action to interdict Southeast Asian sea-lanes. A Chinese attempt to disrupt shipping, for example, would probably: (1) elicit severe ASEAN, regional, and international condemnation and, in particular, deal a severe setback to Chinese efforts to improve relations with the ASEAN countries; and (2) provoke some countries and organizations (e.g., the United States, Japan, the European Union [EU], and ASEAN) to impose economic sanctions, including reductions in investment, trade, and technology transfer. The United States, Japan, and the EU could also block credits to China by international financial institutions.

Third, even if economic and political disincentives failed to deter Chinese military actions to disrupt the sea-lanes, Beijing would need to take into account military, operational, and geographic constraints that would make operations to achieve a closure of the sea-lanes and maritime chokepoints exceptionally difficult. The weaknesses in China's conventional power projection capabilities are detailed in other RAND studies.[3] The key points are summarized below:

- China faces serious shortcomings in its ability to project and sustain force in the South China Sea; in particular, the Chinese navy remains vulnerable to air and surface naval attack. In addition, Chinese forces suffer from low readiness, inadequate training, and deficiencies in logistics support; command, control, communications, computing, and intelligence (C^4I); and modern equipment.

- The Chinese would face serious constraints on the use of mines for interdiction. Although, because of its physical characteristics, the Strait of Malacca is especially vulnerable to mining, the same

---

[3]See, especially, Khalilzad et al.

is not true for moored or bottom mines in the other straits (Sunda and Lombok) and the South China Sea. Hence, even if mining of the Strait of Malacca shut down ship traffic until the channels were cleared, traffic could be rerouted, albeit at additional expense, through the other straits. In addition, the Chinese would have great difficulty in laying mines that can discriminate between enemy and friendly shipping. Further, any overt Chinese mine-laying operation, and in particular reseeding operations, would be highly vulnerable to counterattack and the growing mine-countermeasures capabilities of the ASEAN states. Together, these factors suggest that any blockage of the straits resulting from mines would be either ineffective or limited in duration.[4]

Although China is far from having the across-the-board military capabilities that it would need to challenge the United States or a U.S.-led coalition in the South China Sea, it possesses "asymmetric" capabilities that it could target against specific weaknesses of potential adversaries. China, for instance, has made a substantial investment in modernization of its subsurface naval force. China's Huludao shipyard is the only facility in the Asia-Pacific area building nuclear-powered submarines.[5] Because the South China Sea sound environment is unfavorable to antisubmarine warfare (ASW), Chinese submarines can operate with reasonable effectiveness, despite other operational weaknesses. Chinese short-to-medium-range ballistic and cruise missile systems could also pose a threat to civilian and military shipping.[6]

In short, for at least the next decade China will likely have neither the motivation nor the capabilities to sustain a prolonged closure of the sea-lanes. The Chinese would be inhibited from threatening freedom of navigation in Southeast Asian waters because of the likelihood, if the provocation were great enough, of a severe military

---

[4]Kenny, pp. 23–24.

[5]The first model of a design based on the Russian Victor III is expected to be completed in 2001. See *Jane's Defence Weekly*, February 18, 1998, p. 37.

[6]Evan A. Feigenbaum, "China's Military Posture and the New Geopolitics," *Survival*, Vol. 41, No. 2, Summer 1999, pp. 82–83.

reaction by the United States and other like-minded countries.[7] In this contingency, Southeast Asian states are likely to provide support for U.S. military operations, given the economic and geopolitical consequences they would suffer from a Chinese stranglehold on their economic lifeline. The Chinese would have to calculate, therefore, that any unprovoked attempt to interfere with shipping in the South China Sea would result in a military loss and perhaps a significant increase in the U.S. military presence in China's own backyard.

Conflict over Territorial Claims to the Spratly Islands

Although the prospects are remote that China will mount conventional military attacks against the sea-lanes for the foreseeable future, the possibility cannot be ruled out that hostilities could break out between China and one of the ASEAN states in the South China Sea, perhaps as a result of an incident that spins out of control. In this scenario, China might seek to deter U.S. military involvement by raising the costs of conflict enough to weaken U.S. resolve. The Chinese could calculate, whether correctly or not, that the United States might hesitate to place its carriers at risk, and that China's growing cruise and ballistic missile capabilities would provide Beijing with a credible "sea denial" option.[8]

Indeed, territorial disputes in the South China Sea have emerged as the key external security issue facing ASEAN and pose the greatest potential "flashpoint" for conflict in Southeast Asia (see Figure 3.1). Beijing's quest for improved power projection capabilities, assertiveness in pressing its maritime and territorial claims in the South China Sea, and track record in using force to defend China's sovereignty have all stirred apprehensions in Southeast Asia about China's intentions. Much of the worry reflects an underlying, if often unspoken, fear that Chinese assertiveness foreshadows a China that will become more menacing as its power grows.

---

[7] Kenny, p. 20.

[8] Robyn Lim, "The ASEAN Regional Forum: Building on Sand," *Contemporary Southeast Asia*, Vol. 20, No. 2, August 1998, p. 127.

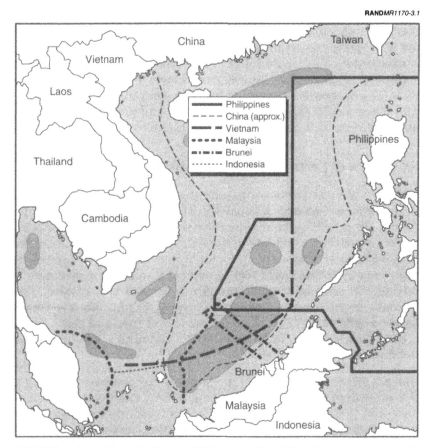

SOURCE: Adapted from http://www.middlebury.edu/SouthChinaSea/maps/oilclaims.gif.

Figure 3.1—Conflicting Claims in the South China Sea

Disputes over the ownership of the Spratly Islands date from a
century of competition among European colonial powers, Japan, and
China for control of the South China Sea. At the present time, the
PRC, Taiwan, Vietnam, Malaysia, Brunei, and the Philippines claim
overlapping parts of the Spratly archipelago and adjacent waters.
The PRC, Taiwan, and Vietnam claim the entire area because
of asserted historical rights. The Chinese base their claim to

sovereignty on the discovery and exploration of the South China Sea by Chinese traders and explorers going back to the second millennium B.C. China's earliest formal claim was made in 1887. Vietnam derives its claim from the jurisdiction exercised by Vietnamese emperors in the early 1800s and rights inherited from the French colony of Cochinchina. Malaysia, Brunei, and the Philippines base their claims on their interpretation of the United Nations Convention on the Law of the Sea (UNCLOS). The Philippine claim is also partially based on the occupation of allegedly unclaimed islands by a private Philippine citizen in 1956. Indonesia is not a party to the Spratly Islands dispute, but the Chinese claims impinge on Indonesian-claimed waters near the Natuna Islands.[9]

Prospects are dim for a negotiated settlement of the Spratly Islands dispute anytime soon, largely because of the wide gap among the ASEAN disputants, the technical complexity of the issues, and China's uncompromising position on the sovereignty issue. Although China has made some rhetorical and tactical shifts in its position on the Spratlys—including the renunciation of force to settle the dispute and proposals for joint development of resources— China has little incentive to reach a diplomatic settlement. Indeed, as one expert on Southeast Asia has observed, China's notion of a settlement is one that endorses China's claims, and Beijing's definition of "joint development" is foreign participation in the exploitation of China's resources.[10]

China's determination to establish its control over the Spratlys stems from historical, political, and economic motives. Although the geopolitical rationale for upholding Chinese claims to the Spratlys has diminished with the end of the Cold War and Soviet military disengagement from the region, such calculations could resurface if China feared containment or encirclement. It is also possible that some Chinese officials see a military presence in the South China

---

[9]Moreover, both the Philippines and Indonesia, as archipelagic states, enjoy the right under the UNCLOS to draw baselines around the fringes of their outermost islands and claim the waters within these boundaries as territorial waters. See Xavier Furtado, "International Law and the Dispute Over the Spratly Islands: Whither UNCLOS? *Contemporary Southeast Asia*, Vol. 21, No. 3, December 1999, pp. 386–404.

[10]See Leszek Buszynski, "ASEAN Security Dilemmas," *Survival*, Vol. 39, No. 4, Winter 1992–1993, p. 94.

Sea—and, more broadly, China's ability to project military force in the heart of ASEAN and threaten control of the sea-lanes—as essential to achieving its goal of regional hegemony.[11]

The Chinese have been unambiguous in stating their position on the status of the Spratlys. Chinese officials often describe the South China Sea claims as a national sovereignty issue and a matter of both pride and principle on which no compromise is possible.

Chinese scholars and academics with close ties to the Chinese government echo this view. As one said: "Regional countries have occupied China's islands and reefs, carved up its sea areas, and looted its marine resources," adding that China's moves in recent years are a "long-overdue and legitimate action to protect its territorial integrity." If China lost such territory, "the legitimacy of the communist regime would be questioned."[12] Echoing this sentiment, another Chinese academic said: "The Spratly issue is about what is China, and what is China's space."[13] Simply put, any Chinese leader considering compromise on the issue would have to take account of the likely adverse reaction of key domestic audiences.[14]

Another factor animating China's desire to establish control over the Spratlys is its growing appetite for oil and the inability of domestic oil production to meet this demand. According to the most recent projections of the United States Energy Information Administration, by the year 2020 China is projected to import 70 percent of its oil and 50 percent of its gas. Virtually all this oil will transit the South China Sea, and thus any disruption of the flow of oil to China could have a crippling effect on the Chinese economy. Historically, the Chinese have sought to minimize their strategic dependence on other countries, and this ideology of self-reliance is still very much alive.

----

[11]See Shee Poon Kim, "The South China Sea in China's Strategic Thinking," *Contemporary Southeast Asia*, Vol. 19, No. 4, March 1998, p. 382, regarding People's Liberation Army (PLA) navy thinking on the strategic importance of control of South China Sea sea-lanes.

[12]Jie, Chen, "China's Spratly Policy," *Asian Survey*, Vol. 34, No. 10, October 1994.

[13]Breckon, p. 49.

[14]See Sheldon W. Simon, *The Economic Crisis and ASEAN States' Security*, Carlisle Barracks, PA: Strategic Studies Institute, U.S. Army War College, 1998, p. 10.

It is likely, therefore, that Chinese leaders are uncomfortable over the prospect of increasing dependence on foreign oil and will look for ways to lessen China's vulnerability to any disruption. Physical control over the Spratlys would achieve this objective in two ways: first, it would prevent other countries from using the Spratlys to mount an oil interdiction effort. Second, the Chinese claim that there are large oil and gas deposits in the waters surrounding the Spratlys and that exploitation of these deposits would help to redress the projected shortfall between oil production and consumption.[15] However, most Western experts believe that these claims are vastly overstated.[16]

In examining the prospects for armed conflict over the Spratlys, two main scenarios merit consideration:

- A Chinese attack on the outposts of other occupants of the Spratlys. For example, the Chinese could conduct air and naval attacks against the garrisons of other claimants to force their departure, which could lead to efforts by the victims to defend their positions or repossess the islands if their units were evicted. Even under these circumstances, however, a major disruption of shipping is unlikely. Most of the commercial shipping that transits the South China Sea passes along sea-lanes that are over 150 miles from likely areas of dispute. In addition, any conflict over the Spratlys is likely to be limited in duration and scope. The Chinese would quickly overwhelm the garrisons on these islands; further, even if the attacked parties decided to mount a counterattack against newly established Chinese outposts, none of the potential belligerents can sustain major force in the Spratlys for more than a few days. Consequently, any disruption of shipping near the Spratlys would be short and the major sea-lanes would remain open.

---

[15]The Chinese Ministry of Geology and Mineral Resources estimated oil and gas reserves of 17.7 billion tons; Ji Guoxing, Director of the Asia-Pacific Department of the Shanghai Institute for International Studies, put reserves at 10 billion tons of oil and 25 billion cubic meters of gas. Cited in South Asia Analysis Group, "Chinese Assertions of Territorial Claims," January 14, 1999, www.saag.org/paper24.html. These estimates of large oil and gas reserves in the South China Sea, however, have yet to be proven by exploration.

[16]Dr. Karl Jackson suggests that Chinese claims of large oil and gas deposits may be a rationalization for Chinese assertiveness in the South China Sea. Comments to authors, February 2000.

- Conflict triggered by energy exploration or exploitation activity, fisheries disputes, accidents or miscalculations, regional tensions, or provocative actions by one or more parties to the dispute.[17] The likelihood of such a scenario, while difficult to judge, cannot be ruled out, particularly in light of previous incidents between China and Vietnam in disputed oil drilling blocks. Since neither country can sustain military operations in this area for long, any hostilities are likely to be of limited duration. But, as was noted above, the Chinese may not have to sustain operations over a lengthy period to attain their political objectives.

AMBIGUOUS THREATS

For the reasons outlined above, the most likely scenario is not a conventional Chinese military attack on territory or forces of an ASEAN state, but a continuation of the successful "island hopping" salami tactics that have marked previous Chinese attempts to extend their control over disputed islands. The Chinese occupation of Mischief Reef, claimed by the Philippines, is a case in point. China may also resort to more subtle and ambiguous uses of force to fulfill its regional goals and ambitions. The Chinese military, for instance, could engage in selective harassment and intimidation of regional states in the guise of enforcement of Chinese maritime claims, protection of Chinese fishermen, antipiracy or antismuggling operations, or restoring stability in the event of the breakdown of domestic or international order. Piracy has always been epidemic in Southeast Asian waters, but the incidence of cases has increased dramatically since the onset of the financial crisis. In 1999, 160 cases of piracy— 56 percent of all the cases reported worldwide—occurred in Southeast Asia. The majority of the attacks occurred in Indonesian and Philippine territorial waters.[18] China itself is the source of much

---

[17]These risks are described in Ralph A. Cossa, "Security Implications of Conflict in the South China Sea: Exploring Potential Triggers of Conflict," *CSIS PacNet Newsletter*, No. 16, April 17, 1998.

[18]Peter Chalk, "Contemporary Maritime Piracy in Southeast Asia," unpublished manuscript, RAND, 2000. Piracy statistics are published by the International Chamber of Commerce, International Maritime Bureau.

of this activity—reportedly with the acquiescence or participation of local officials and customs, police, and naval personnel.[19]

Hence, the most likely challenge the ASEAN countries and the international community will face is periodic Chinese efforts to "pick off" individual islands or reefs, perhaps under the cover of research expeditions or order-keeping operations that deprive other countries of adequate warning.

The range of opportunities for China to engage in this type of activity in Southeast Asia would expand in an environment of economic hardship and political and social disorder. Weakened ASEAN governments unable to control piracy or prevent attacks on ethnic Chinese communities may present Beijing with targets of opportunity for intervention. One factor that is likely to influence Chinese calculations regarding the use of force is whether ASEAN countries, either individually or collectively—or with the assistance of outside powers—have the military capabilities and political will to mount an effective defense against Chinese threats to regional security.[20]

In sum, China's ability to influence the security environment in Southeast Asia will be shaped by political and economic conditions in China and Southeast Asia, by the ASEAN countries' interaction with China, and by the extent to which U.S. and Chinese interests coincide. That said, China is not the only—or for some Southeast Asian countries even the principal—security concern. In a number of regional states, domestic stability holds a higher order of priority.

IMPACT OF ECONOMIC FUTURES

There are four illustrative sets of economic conditions—each discussed in greater detail in the appendix—that could influence the evolution of the Chinese military threat to Southeast Asia: (1) There is no second round of the Asian financial crisis. The region's economies begin to grow again, but at a slower pace than in the pre-

---

[19]Ian James Storey, "Creeping Assertiveness: China, the Philippines, and the South China Sea Dispute," *Contemporary Southeast Asia*, Vol. 21, No. 1, April 1999, p. 100.

[20]Some Southeast Asia specialists take the view that ASEAN states will be unable to mount an effective defense against China, regardless of capabilities and political will, as long as China remains a unitary state.

ceding decade. In this scenario, Asian policymakers' attention will be focused on domestic concerns. (2) There is further economic deterioration. Fragile recoveries are aborted and financial uncertainty again sweeps the region. (3) China manages to get its house in order, but other Asian economies fail to mount sustained recoveries. Within Southeast Asia, there is greater differentiation in economic performance, depending on the individual countries' ability to maintain political stability and appropriate economic and fiscal policies. (4) The Southeast Asian economies begin to recover and resume higher rates of economic growth, but China fails to deal with the structural problems in its banking and state industrial systems.

If we assume no major discontinuities—if the conditions described in illustrative scenarios 1 or 3 apply—the most likely projection is the continuation of Beijing's policy to improve political and economic relations with ASEAN states while exploiting opportunities to strengthen its presence in the region. In these scenarios, China would not directly threaten the territorial integrity of ASEAN states, although a rising China would likely seek to exercise greater influence over ASEAN economic and political policies and shape the regional environment to further its security interests.

If conditions in Southeast Asia deteriorate, China would find more opportunities to pressure or coerce regional countries into acquiescing in its security agenda. China could seek to disrupt ASEAN defense relationships with the United States and other powers, move more aggressively to enforce its territorial claims in the South China Sea, or intervene in domestic conflicts of regional countries, ostensibly to protect local Chinese ethnic communities.[21] If the economic crisis engulfs China as well, the consequences would be unpredictable. China could turn inward, or it could seek to divert attention from domestic problems by ratcheting up international disputes.

---

[21]For a discussion of the relationship of the overseas Chinese to China, see David S.G. Goodman, "Are Asia's 'Ethnic Chinese' a Regional Security Threat?" *Survival*, Vol. 39, No. 4, Winter 1997–1998, pp. 140–155.

CONCLUSIONS

It can be argued that the Chinese seek a peaceful and stable environment in which to promote the expansion of trade and investment. From this perspective, any disruption in its broad patterns of international trade and investment could seriously damage China's ability to sustain high rates of economic growth, which are key to its emergence as a great power and to the preservation of domestic political stability.[22] According to this interpretation, China's concerted diplomatic efforts over the past decade to improve relations with its neighbors, in particular Indonesia and Vietnam, are shaped largely, but not exclusively, by economic considerations.

On the other hand, Chinese military actions ostensibly aimed at preserving regional order and stability may not be inconsistent with China's economic and trade interests. In any event, China's willingness to use force to achieve its objectives is seen by many ASEAN countries as a growing threat to regional stability. Chinese statements and actions clearly reveal growing maritime aspirations in Southeast Asia and a heightened interest in the natural resources of the South China Sea. China's embrace of a more outward-oriented military doctrine, the nature of its military modernization program, and its adventurism in the South China Sea have aroused widespread anxieties about Chinese motivations.

Bellicose Chinese statements about intentions in the region have also fueled perceptions that China's campaign to gain control over the Spratly Islands is part of a larger Chinese expansionist strategy to achieve regional hegemony. In the near to midterm, these incentives and constraints on Chinese behavior suggest that, barring aggressive actions by other claimants, China will act cautiously to press its claims, opting to take advantage of opportunities to achieve quick and easy gains but avoiding truly provocative actions that might precipitate large-scale hostilities or undermine broader Chinese political or economic interests.

---

[22]Although there are dissenters, this view represents a broad consensus among China-watchers. For a recent exposition of this perspective, see Avery Goldstein, "Great Expectations: Interpreting China's Arrival," *International Security*, Vol. 22, No. 3, Winter 1997–1998, pp. 36–73.

Whatever China's long-term intentions toward Southeast Asia, there are the separate issues of whether China would have the military capabilities to prevail in a conventional conflict or whether it might be deterred from using force because of the potential risks and costs that might attend Chinese aggression. In the short to medium term, a conventional Chinese military attack on the territory or forces of an ASEAN state or an attempt to interfere with freedom of navigation on the South China Sea or take control of the Spratly Islands is not the most likely scenario. Given the shortcomings in China's force projection capabilities, the possibility of a quick and decisive victory in a conventional conflict would be far from certain. Unless China felt a sense of encirclement, therefore, military weakness and the political and economic downsides of premeditated aggression are likely to deter China from undertaking large-scale aggression in the South China Sea.

Rather than confronting a conventional military attack, the United States and ASEAN countries are likely to face a continuation of China's creeping irredentism. The challenge will be to devise an effective security strategy to respond to ambiguous moves.[23] For the United States, elements of an effective response might include fostering closer inter-ASEAN defense cooperation, establishing a regional air surveillance network to combat drug smuggling and piracy, and developing the presence and capabilities required to participate effectively in stability-support operations. Implications for U.S. strategy and defense planning are discussed at greater length in Chapter Eight.

---

[23]For an exposition of this view, see Phillip C. Saunders, "A 'Virtual Alliance' for Asian Security," *Orbis*, Spring 1999.

Chapter Four

ASEAN PERCEPTIONS OF A RISING CHINA

Southeast Asian perceptions of China are shaped by the power differential between China and ASEAN, expansive Chinese claims to the South China Sea, the growth in Chinese power projection capabilities, Chinese support for insurgent movements in Southeast Asia in the 1960s and 1970s, the Chinese conflict with Vietnam following the Vietnamese occupation of Cambodia in 1979, and fear of Beijing's manipulation of ethnic Chinese communities in some regional countries. Suspicion of China has been strongest in Indonesia, where there is a long history of tension over the role of the ethnic Chinese community, and in the Philippines, which experienced a still-smoldering Communist-backed insurgency and military clashes with China over disputed areas in the South China Sea. It exists to some extent in all Southeast Asian countries, except possibly in Burma, which maintains a close relationship with China (although there are some recent indications of Burmese wariness), and Thailand, where concerns about China historically have been overshadowed by fear of Vietnamese expansionism.

Together with domestic constraints on intra-ASEAN defense cooperation and lingering intraregional tensions and territorial disputes, the absence of a common perception of an external Chinese threat is a serious impediment, at least in the short to medium term, to effective multilateral defense cooperation among the ASEAN states to deter or confront Chinese military threats. Whether this trend is reversed over the long term will depend primarily on Chinese military assertiveness in the region, perceptions of the credibility of U.S. security commitments and ability to sustain a military presence in the region, the willingness of other powers to participate in security-building coalitions, and the ability of the ASEAN states themselves to overcome their differences.[1] The role of Japan will be of particular note. Although constrained for constitutional and historical reasons from direct military involvement in Southeast Asia, Japan is a strong contender for economic and political influence in the region. Recent Japanese proposals to expand cooperation between the Japanese Coast Guard and regional navies to combat piracy in Southeast Asian waters are indicative of this trend.

Ever since its birth in 1967, ASEAN has viewed China with apprehension. Indeed, the fear that communist governments, underwritten by China, would come to power throughout Indochina was an important catalyst in the decision to form ASEAN and in subsequent efforts to expand regional security cooperation.[2]

Broadly speaking, China's relations with the ASEAN countries have evolved in three phases—which corresponded in turn to changes in the broader security environment:

- **Hostility.** Throughout the 1960s and for much of the 1970s, the ASEAN states had uneasy relations with China. Fear and suspicion of China was pervasive, largely because of Chinese support for communist insurgencies in various ASEAN countries and Beijing's relationship with the large ethnic Chinese communities throughout the region. There were serious differences, in particular, between China and Indonesia and Malaysia, stemming from China's involvement in the 1965 coup in Indonesia and its support for the ethnic Chinese–dominated Malaysian communist insurgency.

- **Thaw.** Beginning in the 1980s, a thaw emerged in China's relations with ASEAN, largely as a result of growing trade and in-

---

[1]These differences were reflected in Malaysia's decision to scale back its participation in military exercises under the Five Power Defense Arrangement (FPDA), the only multilateral arrangement for defense cooperation in Southeast Asia. Tensions between Malaysia and Singapore reportedly contributed, along with Malaysia's economic difficulties, to this decision. See Michael Richardson, "5-Power Defense Pact Is Caught in Crossfire," *International Herald Tribune*, September 22, 1998; and Ian Stewart, "Airspace Ban on Singapore Jet Fighters," *South China Morning Post*, September 18, 1998.

[2]See Michael Leifer, "Expanding Horizons in Southeast Asia?" *Southeast Asian Affairs*, 1994, pp. 3–21.

vestment links spurred by China's economic reforms, the declining role of ideology in Beijing's policies toward the region, China's drive to improve relations with the ASEAN states and its decision to end "dual-track" diplomacy in favor of formal diplomatic relations, and ASEAN's support for China's role as a counterweight to Vietnam. Beijing's efforts to forge closer economic and political relations with ASEAN countries culminated in the 1990–1991 decisions by Indonesia, Singapore, and Brunei to restore official diplomatic relations with China.

- **Ambivalence.** Since the late 1980s, a more ambivalent attitude toward China has emerged. On the one hand, relations among all the ASEAN states and China improved markedly. The growing accommodation between China and ASEAN reflects a mutual desire to achieve economic growth through expanded trade and investment, as well as a conviction that China is destined to become the next East Asian great power and that the most effective way of dealing with Chinese power is to foster greater economic interdependence.[3] China has used the economic crisis to forge closer relations with states that have historically been suspicious of China.[4] At the same time, ASEAN's views of China are not monolithic. Some are more apprehensive of Chinese intentions than others, although in their public posture, these differences are muted by ASEAN's emphasis on consensus.

SINGAPORE

As a small, predominantly ethnic Chinese island-state in the proverbial sea of Malays and dependent on international trade for its economic survival, Singapore reflects a strong dose of *realpolitik*. Of fundamental importance to Singapore is maintenance of the regional and subregional balances of power. In this regard, Singapore's

---

[3]The notion that growing economic interdependence will temper Chinese assertiveness has gained widespread currency, especially among the ruling elites of ASEAN countries. For a typical expression of this view from a Southeast Asian perspective, see Jose T. Almonte, "Ensuring Security the ASEAN Way," *Survival*, Vol. 39, No. 4, Winter 1997–1998, pp. 80–92. Almonte was presidential security adviser and director-general of the National Security Council of the Philippines during the Ramos administration.

[4]"Imperial Intrigue," *Far Eastern Economic Review*, Vol. 160, No. 37, pp. 14–15, September 11, 1997.

overarching concerns are: (1) management of the tensions in the vital relationship with Malaysia and, in the worst-case scenarios, deterrence or defense against Malaysia; (2) political stability in the key neighboring states, especially Indonesia; (3) the future of the U.S. presence; (4) China's long-term intentions; and (5) the evolution of the balance of power in Northeast Asia and the potential for the remilitarization of Japan.[5]

Singapore has several strong reasons to maintain good relations with China. It is one of the largest investors in China and has developed close economic links, including the development of an industrial township in Suzhou at the cost of several hundred million dollars.[6] And unlike other regional states, Singapore has no territorial or maritime disputes with China. Moreover, as a small, vulnerable city-state, Singapore cannot afford to make enemies, particularly a major rising regional power like China. Finally, since almost 80 percent of its population is ethnic Chinese (albeit of different linguistic groups), there are cultural and social affinities that have to be factored into Singapore's policy toward China—although these affinities do not necessarily carry into state-to-state relations.

On the other hand, Singaporeans fear, as do their neighbors, the long-term threat that a hegemonic China could pose to Singapore's independence and look to the United States as the indispensable "balancing" power. At the same time, Singapore has sought to develop links to other powers with a stake in strategic stability, such as Australia and the United Kingdom. Together, these factors have led to a policy that combines economic engagement with China with closer security ties to the United States and other status quo powers. The Singaporeans have also struck a balance between Beijing and Taipei, maintaining strong commercial and informal political ties with Taiwan, while advising Taipei against actions that might precipitate a PRC military response.

---

[5]In discussions with one of the authors, Singaporean international security experts stated that in their thinking ASEAN security included the evolution of the situation in Northeast Asia, particularly on the Korean peninsula. In their view, Russia had disappeared as a factor in Southeast Asian security, but the agreements on peace cooperation and the arms supply relationship with China put Russia on the Chinese side.

[6]The economic return on the Singaporean investment in Suzhou has been disappointing, and the Singaporeans reportedly are cutting back.

Beyond the balancing role of the United States, the Singaporeans see a strong coincidence of interests with the United States, including the maintenance of freedom of navigation, access to regional markets, and global financial stability. Singapore and the United States also cooperated closely in dealing with the consequences of the regional economic crisis and the political crisis in Indonesia. The U.S. relationship is central to Singapore's strategy of strengthening defense technology linkages. Access to U.S. technology, the main source of innovation in defense and information technologies, is critical to the goal of keeping the Singaporean armed forces on the technological cutting edge.[7]

In accordance with this outlook, Singapore has sought to anchor the U.S. military firmly in the region. Singapore hosts the U.S. Navy Logistic Group West Pacific (relocated from Subic in the Philippines) and the USAF 497th Combat Training Squadron, and is constructing berthing facilities to accommodate U.S. aircraft carriers. Singaporean defense cooperation, however, is based on the expectation that the United States remains committed to maintaining a presence in the region. Should confidence in the U.S. commitment falter, Singapore could see no alternative but to accommodate Chinese regional hegemony.

PHILIPPINES

Philippine defense officials and security experts view Chinese expansionism in the South China Sea as the main long-term security threat to the Philippines. The dispute centers on about 50 small islands and reefs in the Spratly Islands, known to the Filipinos as the Kalayaans, some 230 nautical miles west of Palawan.[8] The islands may contain modest potential for gas and oil, but some Philippine officials consider that the fisheries in the disputed areas will provide critical future food security.[9]

---

[7]For the role of defense technologies in Singapore's defense strategy, see Republic of Singapore, Ministry of Defence, *Defending Singapore in the 21st Century*, 2000.

[8]Ian James Storey, pp. 96–97.

[9]Discussion with Secretary of Defense Orlando Mercado, Manila, November 18, 1999.

In 1995, the Chinese occupied an outpost on Mischief Reef, only 150 miles from the Philippines and well within the Philippines' Exclusive Economic Zone (EEZ). The Philippine military's weakness made diplomacy the only realistic option for Manila. Later in the year, the Philippines and China signed a code of conduct aimed at preventing similar incidents in the future. The code provided that no more facilities were to be built or upgraded and that each side would inform the other of naval movements in the disputed area. Nevertheless, in 1997 PLA navy vessels were sighted near Mischief Reef, and in 1998, the Chinese began upgrading the structures, claimed to be fishing shelters, into multistory buildings on concrete platforms, large enough to serve as landing pads for helicopters and manned by Chinese military personnel.[10]

Although the United States has made it clear that it takes no sides on the dispute over the Spratlys, the withdrawal of the U.S. Navy from Subic Bay was no doubt a factor in the Chinese decision to occupy Mischief Reef. In any event, the episode drove home to Philippine decisionmakers the need to revitalize the security relationship with the United States, which had been severely damaged by the failed effort to conclude a new base agreement.[11] In line with the new thinking in Philippine defense policy, the Ramos government negotiated and signed a Visiting Forces Agreement (VFA), the functional equivalent of a status of forces agreement, which would permit the resumption of cooperative military activities with the United States. The VFA was endorsed by the government of Joseph Estrada (a former base opponent), and ratified by the Philippine Senate at the end of May 1999. The VFA was a major step in the reconstruction of the U.S.-Philippine defense relationship and, from Manila's standpoint, of reestablishing deterrence in the region.

Prior to the U.S. withdrawal from its bases, the Philippines relied on the United States to maintain its external security. Therefore, the Philippine armed forces entered the post-U.S.-bases era badly in

---

[10]Storey, pp. 98–107.

[11]The U.S.-Philippine Mutual Defense Treaty applies only to the metropolitan territory of the Philippines, but it calls for bilateral consultations in the event of an attack on the Philippine armed forces. A senior Philippine military officer told one of the authors during a visit to Manila in November 1999 that China would never have dared to occupy Mischief Reef if the U.S. Navy had still been at Subic Bay.

need of modernization. After decades of defense efforts concentrated on the internal communist and separatist threats, the Mischief Reef incident galvanized the Philippines into launching a long-term modernization plan focused on capabilities (e.g., corvettes, offshore patrol vessels, and combat aircraft) that would allow the nation to better defend its claims in the Spratlys and its 200-mile EEZ. Unlike other ASEAN states, the Philippines' economic ties with China are relatively modest and therefore less of an inhibition on Manila's willingness to confront China over its aggressive behavior in the South China Sea.

THAILAND

Thailand has developed strong economic and security ties with China. Thailand's cultivation of a close security relationship with China reflects in part a long-standing Thai tradition of accommodating the region's dominant power.[12] Both countries worked closely together in opposing Vietnam's invasion of Cambodia in 1979 and supported the Khmer resistance to the Vietnamese-installed regime. In addition, the Chinese supplied military equipment to Thailand at bargain-basement prices, including T-69 main battle tanks and naval vessels.

While the treaty-based defense relationship with the United States remains the mainstay of Thailand's security, the economic crisis that struck Thailand and the region in July 1997 presented Beijing with an opportunity to score points with the Thais.[13] Beijing's offer of a $1.0 billion bilateral loan in parallel with the International Monetary Fund (IMF) rescue package contrasted with the U.S. refusal to provide bilateral aid.

Despite Thailand's geopolitical orientation toward China, the absence of any outstanding territorial or maritime dispute, and the ef-

---

[12]Tim Huxley, *Insecurity in the ASEAN Region*, London: Royal United Services Institute for Defence Studies, 1993, p. 35.

[13]Some Thai security analysts worry that since the end of the Cold War there is no longer a common threat that binds the U.S.-Thai defense relationship together. The challenge is how to develop a close relationship in the absence of a common threat. Discussion with Professor Surachart Bamrungsuk, Chulalongkorn University, Bangkok, November 1999.

forts of the Beijing leadership to strengthen relations, the Thai-Chinese bilateral relationship has weakened somewhat in recent years for several reasons. First, Vietnam's withdrawal from Cambodia in 1989 and its entry into ASEAN in 1995, combined with Russia's reduced profile in Asia, have diminished the strategic rationale for Thai-Chinese security cooperation. Second, Thai apprehensions about Chinese intentions and military capabilities in the South China Sea are growing. The Thais are concerned in particular about China's expanding military ties with Burma (Myanmar) and Chinese use of Burmese facilities on the Indian Ocean. Third, the Thais are seeking to develop improved relations with Vietnam, their new partner in ASEAN; hence, the anti-Vietnamese orientation that cemented Thai-Chinese security cooperation has weakened.[14] Nevertheless, China and Thailand recently concluded a joint agreement that could pave the way for a significant expansion of military and economic cooperation.[15]

MALAYSIA

Many Malaysians, especially among the Malay elite and the military establishment, continue to harbor deep suspicion of China's long-term intentions, to some extent because of the bitter historical legacy of China's support for the predominantly ethnic Chinese Malaysian guerrillas during the communist insurgency of the 1950s and 1960s. Moreover, in light of Malaysia's claims to the Spratlys, Malaysians regard their country as a frontline state in the South China Sea dispute. Malaysia's South China Sea policy in recent years, however, has been far from firm in confronting Chinese incursions. Some security analysts in the region believe that Malaysia has decided to accommodate China on South China Sea issues and is emulating Chinese tactics vis-à-vis the Philippines.[16]

---

[14]See Panitan Wattanayagorn, "Thailand: The Elite's Shifting Conception of Security," in Muthiah Alagappa (ed.), *Asian Security Practice: Material and Ideational Influences,* Stanford: Stanford University Press, 1998, pp. 417–444.

[15]Joint Statement of the Kingdom of Thailand and the People's Republic of China on a Plan of Action for the 21st Century, February 5, 1999, press release of the Royal Thai Embassy, Washington, D.C.

[16]Discussions with Singaporean security analysts, Singapore, March 2000.

At a deeper level, Malaysian attitudes toward China are influenced by the interplay of ethnic politics at the core of the Malaysian political system. Although Malaysia has been governed since independence from Great Britain in 1957 by Malay-dominated coalitions of parties representing all of the major ethnic groups, the politics of the Malay majority has been driven by the Malays' fear of losing their dominant position in the state. Singapore's forced separation from the Malaysian Federation in 1965 was an outcome of this dynamic, as were the Kuala Lumpur race riots of May 1969. The political and economic power-sharing arrangements in place for the last 30 years satisfied the Malays' demand for political control and the preservation of their special privileges and gave the Chinese and Indian minorities a role as junior partners in the governing coalition. Together with the high rates of economic growth Malaysia experienced during much of this period, these arrangements have prevented a recurrence of ethnic strife. A protracted economic contraction, however, or a split within the Malay community, could lead to the recurrence of ethnic tensions that might spill over into Malaysian-Chinese and Malaysian-Singaporean relations.

The economic dimension of the Malaysian-Chinese relationship increasingly has shaped Malaysia's attitudes toward China. China's largest overseas investment, a $1.5 billion pulp and paper plant, is to be located in the Malaysian state of Sabah. In Kuala Lumpur's view, exploiting opportunities arising from China's economic modernization and higher political profile could help countries like Malaysia develop leverage vis-à-vis an interventionist West (the United States in particular) seeking to impose its values on Southeast Asian states.[17] A statement by Prime Minister Mahathir is typical of the pragmatic and opportunistic streak in Malaysian policy toward China:

> There is a lot of benefit to be derived from the linkages and the friendship of Malaysian and Chinese peoples. Today, Malaysians

---

[17]See Rosemary Foot, "Thinking Globally from a Regional Perspective: Chinese, Indonesian, and Malaysian Reflections on the Post–Cold War Era," *Contemporary Southeast Asia*, Vol. 18, No. 1, June 1996, pp. 20–21.

are investing and helping China to develop. The past is very much forgotten and in many ways irrelevant.[18]

Chinese leaders, in turn, have voiced support of Mahathir's attacks on international financial circles that Mahathir blames for the Asian financial crisis. During then-Premier Li Peng's visit to Malaysia in September 1997, Li and Mahathir agreed that other centers of power should be developed in Europe and Asia to balance U.S. predominance.[19] Malaysia's rapprochement with China paid dividends for Beijing by helping to block the emergence of an ASEAN consensus in opposition to China's claims in the South China Sea.[20]

INDONESIA

Deep-seated Indonesian suspicions of China have been submerged, for the present, by preoccupation with domestic instability and the new foreign policy direction of the Wahid government. President Wahid has been seeking to improve ties with China, and spoke of a Beijing-New Delhi-Jakarta "axis" (an unfortunate term that has since been downplayed in official discourse). Suspicion of China, however, remains strong among the Indonesian elite and the military. This suspicion stems from Beijing's involvement with the Indonesian Communist Party in the abortive 1965 coup and continued fears that Beijing might seek to manipulate domestic Indonesian politics.[21] Indonesians are also wary of China's intentions in the South China Sea. Although Indonesia is not a claimant in the Spratlys dispute, Jakarta's fears of China have been kindled by China's claims to sovereignty over the entire South China Sea and by China's continuing buildup of power projection capabilities. Indonesian fears of Chinese ambitions were exacerbated by the publication of a Chinese

---

[18]Cited in Derek Da Cunha, "Southeast Asian Perceptions of China's Future Security Role in Its Backyard," in Jonathan D. Pollack and Richard H. Yang (eds.), *In China's Shadow*, RAND CF-137-CAPP, 1998, p. 115.

[19]See Foreign Minister Abdullah Badawi's statement, cited in *Far Eastern Economic Review*, September 11, 1997, p. 15.

[20]Malaysia's attitude at the Hanoi ASEAN Summit in December 1998 scuttled any chance of a joint ASEAN acknowledgment of China's expansion of its foothold on Mischief Reef, in an area claimed by the Philippines. Dr. James Clad's comments to authors, January 2000.

[21]Discussions with Indonesian military and security experts, Jakarta, March 2000.

map that identified part of the waters off the Indonesian island of Natuna, a major natural gas field, as Chinese territorial waters.[22] The Jakarta government manifested its concern that China's assertiveness might challenge strategic Indonesian interests in the area by holding air and sea exercises in December 1995 off the Natuna Islands, to which Jakarta-based defense attachés, including the PRC attaché, were invited.

Indonesian concerns about China's intentions do not necessarily portend, however, closer Indonesian military ties with other states or a more confrontational military stance toward China. Indonesian defense policy remains preoccupied with threats to the country's unity and stability. The armed forces see external threats as remote, and the Indonesians do not perceive the "Chinese threat" in conventional military terms but rather in terms of Chinese attempts to exploit Indonesia's lack of political and social cohesion.[23] These threat perceptions are reflected in Indonesia's low level of military expenditures and the modest pace of the Indonesian defense modernization program, even before the onset of the economic crisis.

Throughout most of Indonesia's history as an independent state, its concept of security, often referred to as "national resilience," and the associated military doctrine of "total defense and security," stress self-reliance in defense and national economic and social development to contain internal threats to national unity and stability. This security concept and military doctrine are only now beginning to change with the separation of the police from the armed forces and the transfer of internal security functions to the police.[24]

---

[22]The Chinese map's claim to the Natuna waters was repeatedly brought up by Indonesian security analysts associated with think tanks of different political persuasions during one of the authors' trips to Jakarta in November 1997. As an archipelagic state, Indonesia, like the Philippines, pursuant to Part IV of UN Convention on the Law of the Seas, claims as territory all waters within a baseline defined by its outer islands.

[23]Dewi Fortuna Anwar, "Indonesia: Domestic Priorities Define National Security," in Alagappa, pp. 477–512.

[24]Presentations by A. Hasnan Habib and LTG TNI Agus Widjojo at the Council for Security Cooperation in the Asia Pacific (CSCAP) Conference on Indonesia's Future Challenges, Jakarta, March 8–9, 2000.

The Indonesian government and military have been preoccupied by the insurgencies in East Timor (until its de facto separation), Aceh, and West Papua (formerly Irian Jaya), the unrest in Riau, and religious and ethnic clashes in the Moluccas, Sulawesi, and Kalimantan. The growing demands of the outlying islands for greater autonomy from Jakarta may create greater stresses on the Indonesian political system. The Indonesian armed and security forces are too thinly stretched to confront these challenges and to keep order in Java as well, should there be an upsurge in political turmoil.

From Jakarta's perspective, therefore, any serious downturn in Chinese-Indonesian relations or an escalation in military competition and tensions between the two countries could compromise the central government's primary goal of maintaining internal order. Hence, Jakarta's perception of a Chinese military threat is likely to be tempered by these internal security considerations.

VIETNAM

Despite normalization of relations and expanding bilateral trade, Vietnam continues to see China as an external threat and remains suspicious of China's intentions and ambitions. China casts a large shadow over Hanoi's strategic outlook for several reasons.

First, historical memories of Chinese domination, invasions, and border conflicts have engendered a deep and abiding mistrust of China. Sino-Vietnamese differences were submerged by the intrusion of the Europeans and Japanese and the political and ideological conflicts of the first half of the twentieth century. During the French colonial period, many Vietnamese nationalists had connections with China, and China provided critical aid to the Vietnamese Communists during the Indochina and Vietnam Wars. After Hanoi's conquest of the South in 1975, a series of related developments—Hanoi's tilt toward Moscow in the Sino-Soviet dispute and the anti-Vietnamese orientation of the Khmer Rouge and their alignment with Beijing—led to a collision between Vietnam and China, culminating in the 1979 border war.[25]

The collapse of the Soviet Union provided the impetus for the improvement in relations between Vietnam and China. Relations were normalized in 1991, and in January 2000 a potential source of conflict was removed with the signing of the Land Border Treaty between Vietnam and China. Nevertheless, despite official declarations of amity, tensions are not far from the surface. Although Vietnam accrues economic benefits from the border trade between the two countries, this activity has been accompanied by a large degree of smuggling, crime, and corruption that many Vietnamese attribute to a deliberate Chinese policy of destabilizing Vietnam's domestic market and damaging Vietnamese industries.[26]

Second, Vietnam is a primary protagonist in the Spratly Islands dispute and the two countries have had armed confrontations in 1974 and 1988 over the Paracel Islands and the Spratlys, respectively.[27] In 1992, China occupied the Da Ba Dan and Dac Lac reefs, built oil-drilling platforms in disputed areas of the Gulf of Tonkin, and granted an oil concession to Crestone Energy Corporation, a U.S. energy company, in an area contested by the Vietnamese. In 1997, the Chinese conducted exploratory drilling in what was supposedly Vietnam's continental shelf, and in 1998 it was reported that the Chinese had erected a ground satellite station in the Paracels and a telephone booth in the Spratlys.[28]

The current reconciliation between Vietnam and China thus remains fragile, and further belligerent Chinese actions in the South China Sea could revive Vietnam's fear of China and lead to a more hostile and confrontational posture, including over the long term a desire for closer military relations with the United States. That said, the

---

[25]See Ang Cheng Guan, "Vietnam-China Relations Since the End of the Cold War," Institute of Defence and Strategic Studies (IDSS), IDSS Working Paper, Nanyang Technological University, Singapore, November 1998.

[26]Kim Ninh, "Vietnam: Struggle and Cooperation," in Alagappa (ed.), p. 31.

[27]Unlike Spratlys, where there are multiple overlapping claims, in the Paracels Vietnam is the only ASEAN state contesting China's claims.

[28]Ang, pp. 8–11; Richard Betts, "Vietnam's Strategic Predicament," *Survival,* Vol. 37, No. 3, Autumn 1995.

Vietnamese are keenly aware of their own vulnerabilities vis-à-vis China and remain preoccupied with addressing the country's economic and social development through economic liberalization and increased participation in the global economy (*doi moi*, or "renovation policy"). At least for the moment, Vietnam's strategy for dealing with China emphasizes continued normalization of relations, solidarity and integration with ASEAN, military modernization largely with Russian equipment, and expanding economic and political ties with outside powers, especially the United States, Japan, and EU countries. Secretary of Defense William Cohen's groundbreaking visit to Vietnam in March 2000 should be seen in this context.

Chapter Five

ASEAN DEFENSE POLICIES AND EXPENDITURES

Without a perception of a common threat from China and in the presence of the continuing tensions and disputes among ASEAN countries, intra-ASEAN defense cooperation remains limited. Over the past decade, all ASEAN decisions related to defense expenditures, weapon acquisitions, and force modernization have been made on a national basis without intra-ASEAN coordination, reflecting ASEAN interest in defense cooperation to promote "confidence-building" rather than functional cooperation to achieve a common defense objective.[1] Furthermore, although apprehensions about Chinese intentions have influenced the defense policies and programs of some ASEAN states, many of the ASEAN states' defense expenditures and programs stem primarily from domestic political considerations; intra-ASEAN tensions; the desire to combat piracy, smuggling, and drug trafficking; and the growing interest in monitoring and protecting EEZs and fishing areas.

Over the past few years, several ASEAN countries have developed a network of informal bilateral defense ties that is often described as an "ASEAN defense spider web." Underpinning this form of cooperation is a widespread conviction on the part of ASEAN leaders that bilateral cooperation offers advantages over other forms of multilateral military cooperation. In the words of the former chief of the Malaysian armed forces:

---

[1]Huxley, p. 66.

Bilateral defense cooperation is flexible and provides wide-ranging options. It allows any ASEAN partner to decide the type, time, and scale of aid it requires and can provide. The question of national independence and sovereignty is unaffected by the decision of others as in the case of an alliance where members can evoke the terms of the treaty and interfere in the affairs of another partner.[2]

Within ASEAN, mutual use of facilities has increased and there has been a significant increase in joint military exercises, with a focus on air and naval operations in maritime scenarios. For example:

- The Thai and Singapore air forces train together in the Philippines, and Singapore has also had access to excellent training facilities in Brunei.

- Malaysia and the Philippines have a bilateral defense cooperation agreement that provides for regular joint military exercises, military information exchanges, and the possible use of each other's military facilities for maintenance and repair.

- Singapore cultivated defense ties with Indonesia and reached agreements that allow Singapore to conduct naval exercises in Indonesian waters and to use air combat ranges in Sumatra.

- Under the aegis of the FPDA, Malaysia and Singapore expanded military cooperation to include participation in annual exercises and the organization's Integrated Air Defense System (IADS). Bilateral military cooperation took a turn for the worse after 1998 as a result of political disputes between the two countries. Malaysia pulled out of an FPDA combined exercise (although it later announced it would resume participation) and rescinded agreements that allowed Singaporean military and rescue aircraft to overfly Malaysian territory without prior authorization.

- Malaysian-Thai joint air exercises have been extended to patrol maritime areas.

---

[2]As quoted in Amitav Acharya, "Regional Military-Security Cooperation in the Third World: A Conceptual Analysis of the Relevance and Limitations of ASEAN," *Journal of Peace Research*, Vol. 29, No. 1, 1992, p. 13.

- Indonesia and Malaysia developed close bilateral defense cooperation, including regular military exercises and frequent high-
  level military exchanges and visits.[3]

Many of these bilateral ties, especially those related to intelligence
sharing and enhanced military contacts, are designed to promote
greater transparency and understanding to remove mutual suspicions and tensions, or to combat common security problems in border and maritime areas, including smuggling, drug trafficking,
piracy, and protection of EEZs. However, advances in defense cooperation among ASEAN countries and with extraregional powers suggest a growing interest in defense against external threats. For example, in the early 1990s, in anticipation of the U.S. withdrawal from
bases in the Philippines, Singapore and the United States concluded
agreements that allow U.S. ships and aircraft to use Singapore's military facilities for repair, resupply, and logistics support. This cooperation took a significant step forward with Singapore's decision to
upgrade dock facilities to accommodate visits by U.S. aircraft carriers. The United States has modest logistics support agreements with
Indonesia, Malaysia, and Brunei. In addition, the United States con-
ducts annual military exercises with Thailand, including *Cobra Gold*,
and periodic bilateral military exercises with the Philippines.
Singapore uses military facilities in Australia, Israel, Thailand,
Taiwan, Brunei, and the United States. Australia and Indonesia
concluded a bilateral security agreement in 1995,[4] but Indonesia
renounced the agreement in 1999 to protest Australian criticism of
Indonesia's East Timor policy. The ASEAN countries also have a
variety of defense arrangements with a number of EU countries,
although of lesser significance and often tied to commercial deals.[5]

Despite the limited progress in expanding ASEAN military cooperation, without a major shift in strategic perspectives and deeply in-
grained habits of thinking, prospects are dim in the short to medium

---

[3]Sheldon W. Simon, "The Regionalization of Defense in Southeast Asia," *The Pacific Review*, Vol. 5, No. 2, 1992, p. 119.

[4]Michael Leifer, *The ASEAN Regional Forum*, Adelphi Paper 302, Oxford University Press for IISS, 1996, p. 14.

[5]Paul Stares and Nicolas Regaud make the case for greater European involvement in "Europe's Role in Asia-Pacific Security," *Survival*, Vol. 39, No. 4, Winter 1997–1998, pp. 117–139.

term that ASEAN will evolve into an effective regional collective security or defense organization with coordinated doctrine, training exercises, planning, procurement, weapons production, and interoperability. Even with growing concerns over China's potential threat to regional security—and the recognition that individual ASEAN states are unable to mount a credible defense against China—there are formidable obstacles to multilateral defense cooperation:

- As Barry Buzan and other scholars have noted, ASEAN countries (with the exception of Singapore) are "weak states" characterized by a lack of political and social cohesion.[6] The weakness of these states—reflected in the continuing preoccupation of ASEAN members with internal security and regime survival—makes intra-ASEAN defense and security cooperation more difficult.[7] Moreover, given the differing perceptions of threats from China, individual ASEAN states believe they can fashion a bilateral avoidance strategy that works better than a coalition strategy.[8]

- By and large, ASEAN leaders have manifested an inward orientation on security matters. Their key objective has been the attainment of national or regional "resilience," and many continue to believe that a multilateral military pact or defense alliance is irrelevant and ineffective in meeting the ASEAN states' most serious security requirements.

- There is a widespread belief among ASEAN leaders that any effort to turn the organization into a formal military pact would fracture the cohesion of ASEAN, which has been weakened by ASEAN's expansion and the inclusion of new members with divergent security orientations and threat perceptions.

- ASEAN militaries lack a common doctrine and language, standardization of equipment, and common logistical support infrastructure. Despite the potential operational and financial bene-

---

[6]Huxley, pp. 12–14.

[7]Dana R. Dillon, "Contemporary Security Challenges in Southeast Asia," *Parameters,* Spring 1997, pp. 119–133.

[8]Dr. James Clad's comments, January 2000. As noted in the preceding chapter, this is also the view of Singaporean security analysts with regard to Malaysian policy vis-à-vis China.

fits, ASEAN countries have made little effort to harmonize their weapons procurement or production policies.

• Cultural factors tend to inhibit movement toward meaningful defense cooperation. Some elements of the regional strategic culture that have been noted in this regard include a desire to seek consensus over confrontation; reliance on bilateral rather than multilateral approaches to security planning; an emphasis on informal structures and personal relationships cultivated away from formal meetings; comprehensive approaches to security that stress the economic, social, and political dimensions of national security; and roles for the military that go well beyond national defense.[9]

• Lingering tensions and suspicions and unresolved ethnic and territorial disputes pose a serious impediment to expanded intra-ASEAN defense cooperation. The most important of these involve Thailand's tense relationship with Burma/Myanmar, the Philippines' dispute with Malaysia over the province of Sabah, the competing claims of the Philippines, Malaysia, and Vietnam in the South China Sea, territorial disputes between Malaysia and Indonesia and Malaysia and Thailand, and tension between Singapore and Malaysia dating back to Singapore's forced separation from the Malaysian Federation in the mid-1960s. Indeed, as one ASEAN specialist has noted, Singapore continues to base its defense strategy primarily on deterrence of its larger neighbors and Singapore and Malaysia still plan for war against each other.[10]

The key issue, however, is that threat perceptions of China differ—at one end of the spectrum the Philippines perceives the threat as immediate and is seeking to develop an ASEAN consensus in opposi-

---

[9]See Desmond Ball, "Strategic Culture in the Asia-Pacific Region," *Security Studies*, Vol. 3, No. 1, Autumn 1993, pp. 46–47.

[10]Tim Huxley, "Singapore and Malaysia: A Precarious Balance?" *Pacific Review*, Vol. 4, No. 3, 1991; and Andrew T.H. Tan, "Singapore's Defence: Capabilities, Trends and Implications," *Contemporary Southeast Asia*, Vol. 21, No. 3, December 1999, pp. 453–457.

tion to Chinese assertiveness;[11] others, such as Malaysia, rely on bilateral avoidance strategies. Hence, if China embarks on an expansionist course, the primary responsibility for defense will most likely fall on an ad hoc coalition of willing countries. To the extent that the ASEAN countries closer to the scene and with the most relevant military capabilities participate in this coalition, ASEAN could raise the costs and risks of Chinese aggression and thus deter China from using force. However, recent trends in force development and modernization, defense budgets, and arms procurement do not offer grounds for optimism.

SINGAPORE

Since the early 1970s, Singapore has allocated an average of 6 percent of its GDP to defense expenditures, which has enabled it to acquire, for a state of Singapore's size, very capable, modern, and well-trained ground, air, and naval forces. Moreover, the economic crisis has not had a significant impact on defense spending or force modernization. In fact, the defense budget increased from S$6.1 billion to S$7.3 billion in 1998. Planned defense expenditures also increased, in U.S. dollars, from $4.1 billion to $4.3 billion over the same period.[12] The air force has close to 200 modern aircraft in its inventory, including two squadrons of F-16s, three squadrons of F-5Es reconfigured for maritime strike and reconnaissance missions, three squadrons of upgraded A-4 Super Skyhawks, and eight maritime patrol aircraft. E-2C patrols have been extended well into the South China Sea and these aircraft, if deployed at bases in Malaysia, would be able to loiter in the vicinity of the Spratly Islands for a prolonged period. Moreover, the F-5Es have a midair refueling capability, which extends their range and loitering capability well into the South China Sea.[13] The air force has also taken delivery of a number of Malat Scout remotely piloted vehicles (RPVs) from Israel.[14] The navy has

---

[11]See Jose T. Almonte, "ASEAN Must Speak with One Voice on the South China Sea Issue," paper delivered at the South China Sea Confidence-Building Measures Workshop, Jakarta, March 10–11, 2000.

[12]*The Military Balance 1998/99*, International Institute for Strategic Studies, Oxford: Oxford University Press, p. 195.

[13]Republic of Singapore, pp. 32–33; Simon, 1992, p. 116.

[14]Tan, p. 459.

three squadrons operating six missile corvettes, six missile gunboats, and six antisubmarine-capable patrol craft, armed with Harpoon, Barak, and Mistral missiles and Whitehead torpedoes, and has acquired four Type-A12 submarines from Sweden.[15] The Singaporean armed forces recognize the critical importance of technology and have entered a new phase of military development that emphasizes information, sensing, precision attack, stealth, and aerospace warfare technologies.[16]

PHILIPPINES

Both the Aquino and Ramos administrations backed military modernization programs, but despite the alarm over the PRC encroachment in the Spratly Islands, little progress has been made in upgrading the armed forces' capabilities. The Philippines does not currently have a modern military posture capable of independent defense of its territorial waters and claims in the Spratly group. Most of the armed forces' equipment is obsolescent or suffers from poor readiness. The air force has five airworthy F-5A/Bs.[17] Philippine bases have been in a state of disrepair since the U.S. withdrawal earlier in the decade. Moreover, the country is still plagued by a low-level internal insurgency that drains funds away from upgrading of air and naval forces.

The Philippines increased defense expenditures, in pesos, from P39 billion in 1996 to P42 billion in 1997 and P47 billion in 1998. However, in dollar terms, defense spending fell from $1.5 billion in 1996 to $1.2 billion in 1998.[18] The centerpiece of the Philippine military modernization plan is acquisition of a squadron of advanced

---

[15]Republic of Singapore, pp. 35–36; Singapore acquired its first Type-A2 submarine for training in 1995. Part of a regional trend, in the mid-1990s Indonesia, Thailand, and Malaysia all entered into contracts or requested proposals for submarine acquisitions, but these programs were frozen or cancelled as the result of the economic crisis. A possible impetus for this interest in submarines was the submarine modernization program in the PRC.

[16]Tan, pp. 466–467.

[17]According to military sources, Taiwan offered to transfer F-5s to the Philippines at a nominal cost in exchange for the use of training facilities; the transaction was opposed by the Foreign Ministry because of the Philippine government's One China policy.

[18]*The Military Balance 1998/99*, p. 194.

fighter aircraft and naval combat vessels. The Estrada administration
has reaffirmed its intention to proceed with this ambitious modernization plan, but it remains to be seen whether the government will
be able to implement it, given the uncertain prospects for the resumption of sustained economic growth and the competing demands for social spending.

THAILAND

Thailand's military doctrine has gradually shifted from an emphasis
on small-scale warfare against internal ground threats to a more
outward-looking maritime orientation and balanced conventional
defense posture. Reflecting this shift, modernization of the Royal
Thai air force and navy remains a priority. However, even before
Thailand's current financial crisis, the Thai government had rele-
gated defense programs to a lower priority, as evidenced by the sharp
decline in defense spending as a percentage of GDP and total government spending. From 1985 to 1998, defense expenditures as a
percent of GDP dropped from 5.0 percent to 1.5 percent.[19] As a consequence of the Asian financial crisis, defense expenditures, in Thai
baht, fell from b102 billion in 1997 to planned expenditures of b81.0
billion in 1998 and b77.4 billion in 1999. In dollar terms, this represents a decline from $3.2 billion in 1997 to $1.8 billion in 1999.[20]
According to Thai military sources, in 2000 military expenditures are
expected to rebound to b88.6 billion.

Despite these setbacks, Thailand has been able to continue some
modernization programs. The Royal Thai navy has 14 frigates, 5
corvettes (many of which are armed with Harpoon antiship missiles),
and more than 80 patrol and coastal vessels. Two new classes of
frigates will enter the inventory within a few years, and a light aircraft
carrier (*Principe de Asturias* type), with a complement of eight
Spanish AV-8S Matador (Harrier) and six S-70B Seahawk helicopters,
was commissioned in 1997 and will significantly improve the navy's
power projection capability, although lack of funds has kept it at a
low state of readiness. The Royal Thai air force added 36 F-16s in the

---

[19]Huxley, Tim, and Susan Willett, *Arming East Asia*, Adelphi Papers 329, Oxford:
Oxford University Press, 1999, p. 17.

[20]*The Military Balance 1998/99*, pp. 198–199.

mid-1990s, but was forced to shelve plans to acquire eight F-18s. It is also in the process of upgrading its air defense and electronic surveillance capabilities. Acquisition of an airborne early warning system, perhaps the E-2C Hawkeye, has been indefinitely postponed, as was the Thai navy's submarine program.

MALAYSIA

Since the 1980s, Malaysia has been reorienting its force structure to a posture designed to protect maritime and territorial claims in the South China Sea. The new policy was designed in response to the strategic environment shaped by the end of the communist insurgency, a diminished U.S. military presence in Southeast Asia, and fears of Vietnamese expansionism and Chinese assertiveness.[21] Although both Malaysia and China have avoided clashes over disputed areas in the South China Sea, there is the potential for Chinese occupation of Malaysian-claimed areas in the Spratlys.[22]

Malaysia is in the latter phase of implementing an $8.5 billion defense modernization program launched several years ago, before the onset of the regional economic crisis. Malaysia has close to 95 combat aircraft in its inventory, including 18 MiG-29Ns, 8 F/A-18Ds, 25 BAe Hawk fighter/bombers, and 13 F-5Es. Some of these aircraft can be refueled in midair, and the Malaysian air force trains extensively for maritime operations beyond territorial waters. The Malaysian navy operates 40 frigates, patrol craft, and coastal vessels armed with Seawolf surface-to-surface missiles and Exocet antiship missiles. The armed forces can move a rapid-deployment force and three airmobile battalions to the Spratlys with a combination of C-130s and amphibious craft under navy escort.[23]

In the 1980s, Malaysia developed an air base on Labuan island, in northern Borneo, intended to strengthen the defense of Sabah (and

---

[21]J. N. Mak, "The Modernization of the Malaysian Armed Forces," *Contemporary Southeast Asia,* Vol. 19, No. 1, June 1997, pp. 29–51.

[22]A potential flashpoint is the Layang Layang islet, halfway between the Spratly Islands and the northern coast of Borneo, where Malaysia maintains a small garrison. *Jane's International Defense Review,* September 1997, p. 23.

[23]*Jane's,* September 1997; *The Military Balance 1998/99,* p. 189

perhaps intimidate Brunei, which still fears absorption into Malaysia) and project power into the South China Sea.[24]   The upgrading of the naval base at Sandakan, at a cost of $450 million, was also a high pre-crisis defense priority.[25]

Cuts in the Malaysian defense budget cast doubt on Malaysia's ability to complete its military modernization goal. In 1998 the defense budget decreased by approximately 11 percent in Malaysian ringgit, from RM9.5 billion to RM8.5 billion. This represented a 38 percent decline in dollar terms, from $3.4 billion to $2.1 billion.[26] As a result, several planned procurements have been put on hold.

INDONESIA

Indonesia's conventional defense capability remains modest. In the past 15 years, defense spending as a percentage of GDP has declined from 4.2 percent to 1.5 percent, notwithstanding an average annual GDP growth rate of 5.5 percent during the decade preceding the onset of the financial crisis in 1997. Further, 60 percent of the defense budget is allocated to personnel, and the small procurement budget has often been used to acquire weapons for political or prestige reasons. The navy has 17 main combatants in varying states of seaworthiness and about 100 corvettes and patrol craft— insufficient to maintain security in waters that have been subject to increasing activity by pirates. The air force flies a combination of aircraft, including one squadron of F-16s (10 aircraft) of which about half are airworthy, a squadron of C-130s in similar condition, and Hawk, refurbished A-4, OV-10, and Bronco aircraft.[27] Still, Indonesia intensified its military cooperation with Singapore and Malaysia and, before the economic crisis, planned to strengthen its air force and naval capabilities. However, Indonesia's current economic turmoil has led it to suspend indefinitely its plans to purchase 12 advanced Russian SU-30MK combat aircraft and 8 Mi-17 helicopters and five

---

[24]Dr. James Clad's comments, January 2000.

[25]"Malaysia Strains for a Greater World Standing," *Jane's International Defense Review,* April 1997, p. 25.

[26]*The Military Balance 1998/99,* p. 189

[27]*Jane's International Defense Review,* September 1997, pp. 33–36; discussions with senior Indonesian Air Force officers, Jakarta, March 2000.

Type-209 submarines from Germany. Also, while the defense budget for 1998 increased by approximately 43 percent from 14 trillion to 20 trillion rupiah, the collapse of the rupiah resulted in a decrease, in dollar terms, from $4.8 billion in 1997 to $1.7 billion in 1998.[28]

VIETNAM

Vietnam has witnessed a major military retrenchment over the past several years, fueled in large measure by the cutoff of Soviet aid in 1991, which had underwritten Vietnam's military buildup, the withdrawal of Vietnamese forces from Cambodia, and the goal of economic modernization. As a result, although the Vietnamese continue to view China as a long-term adversary, particularly over the contested Spratly Islands, lack of funds for spare parts, training, and maintenance have taken a toll on the readiness of Vietnam's air and naval forces. Thus, even though the Vietnamese have 200 combat aircraft, including 65 SU-22s and 6 SU-27s, many of these aircraft are not operational and the Vietnamese would have great difficulty operating effectively with other ASEAN forces. The Vietnamese hope to strengthen their air force with an additional 24 SU-27 air superiority/ground attack aircraft. Likewise, Vietnam is seeking naval vessels from Russia.[29] Although the Vietnamese navy has over 60 frigates, patrol craft, and coastal vessels in its inventory, many of these ships are in serious disrepair.

THE SINO-ASEAN POWER IMBALANCE

The gap in military capabilities between the ASEAN countries and China is likely to grow over the next 10 to 15 years. First, at least in the short term, the economic downturn in Southeast Asia is likely to diminish prospects for closer military cooperation among the ASEAN states and for the ability of those states to develop a credible military deterrent against external threats. The serious economic and social dislocations resulting from the economic crisis have turned the attention of governments and armed forces to internal security threats.

---

[28]*The Military Balance 1998/99*, p. 181

[29]*Jane's Defence Weekly*, January 6, 1999, p. 12.

Second, internal economic strains have led to political tensions among several ASEAN states or revived long-standing disputes that had been suppressed by economic prosperity. Malaysia's relations with both Singapore and Indonesia have been strained over refugee, immigration, and other economic issues. Friction between Thailand and Burma over border issues is also on the rise.

Third, because of their economic woes and growing preoccupation with internal security problems, most of the ASEAN states (Singapore is the exception) have slashed defense expenditures, weapons procurement, and force modernization. As a result, there has been a decline in combined exercises and training. Modernization of air and naval forces and other programs to enhance ASEAN force projection capabilities have been delayed, cut back, or canceled. Moreover, because ASEAN states have not coordinated any of these decisions, interoperability within ASEAN, which has traditionally been weak anyway, has been dealt a further setback.

Fourth, the financial crisis undermined ASEAN political solidarity, which historically has underwritten progress in defense cooperation. The impact of the economic crisis on ASEAN's cohesion was amplified by ASEAN's enlargement, which made the organization less homogenous. The fissures within ASEAN were reflected in its tepid response to China's recent military buildups on Mischief Reef in the Spratlys and Woody Island in the Paracels. The ASEAN summit meeting in Hanoi in December 1998 ended in disarray, with major disagreements over the immediate admission of Cambodia, trade, and financial issues. As a result, little further progress was made in fostering greater regional transparency, dialogue, security cooperation, and trust-building. Because of its unsettled domestic situation, Indonesia was unable to assert its traditional role of regional leadership, and it is uncertain that any other ASEAN country has the will or resources to fill this void.[30]

It is too early to tell whether the financial crisis will have a lasting impact on ASEAN political cohesion and defense cooperation. Beijing, as it has done in the past, has taken advantage of ASEAN disarray to strengthen its military positions on disputed islands in the Spratlys.

---

[30]See Anthony Smith, "Indonesia's Role in ASEAN: The End of Leadership?" *Contemporary Southeast Asia*, Vol. 21, No. 2, August 1999.

Before the crisis, ASEAN acquisition of modern weapons over the past 15 years had outstripped China's. Today, this situation has been reversed, leading one prominent observer to conclude, "the timetable for the PLA to catch up with and perhaps surpass its Southeast Asian neighbors may well be accelerated."[31]

At the same time, it may be premature to conclude that the changes precipitated by Southeast Asia's economic downturn will become a permanent fixture of the regional landscape. There are encouraging signs of recovery in Thailand, Malaysia, and even Indonesia. Although economic growth and defense expenditures may not return to precrisis levels, they could rebound sufficiently to sustain moderate growth in defense capabilities. China has taken advantage of ASEAN's distractions to beef up its military capabilities in the South China Sea and to expand its political and economic influence in the region. Nonetheless, the Chinese have refrained from currency devaluation that would have aggravated ASEAN's economic difficulties and have shown restraint in the face of violence against the ethnic Chinese minority in Indonesia.

Finally, many of the intra-ASEAN disputes described above predate the crisis and only rose to the surface after lying dormant for many years because of the suddenness and magnitude of the economic collapse. Indeed, ASEAN states had made some progress in mitigating or containing many of these tensions. Although the current malaise has arrested this trend, it may well resume once the crisis has passed, as long as the current squabbles do not rupture political relations or escalate into military confrontations, either of which could cause lasting damage to the fabric of intra-ASEAN relations.

In sum, the economic crisis has diminished ASEAN security as well as the credibility, effectiveness, and prestige of the "Asian way" of managing regional relations.[32] The military balance between China and ASEAN has shifted in China's favor and ASEAN's capacity to resist Chinese encroachments has diminished. ASEAN's institutional strength and cohesion have weakened, and there are few signs that ASEAN's leaders have a coherent plan for reinvigorating ASEAN soli-

---

[31]Sheldon W. Simon, p. 22.

[32]Amitav Acharya, "A Concert of Asia?" *Survival*, Vol. 41, No. 3, Autumn 1999, pp. 84–99.

darity. Intra-ASEAN security cooperation has come to a virtual halt, and many ASEAN countries are absorbed by internal threats and challenges. All this said, many of these trends and developments are not new and some probably would have occurred even if there had been no economic crisis. Thus, while some of these trends, such as China's enhanced military presence, have been accelerated, they were not created by the crisis.

Chapter Six

REGIONAL APPROACHES TO SECURITY COOPERATION

The ASEAN countries have manifested a marked preference for multilateral approaches to regional security problems. This preference stems primarily from three factors: the proliferation of transnational problems that cannot be solved at the national level, uncertainty about the future of the U.S. security role in the region,[1] and the expectation that locking China into multilateral security arrangements might constrain its behavior and induce it to take greater account of ASEAN interests and sensitivities.[2]

The ASEAN Regional Forum (ARF), first held in July 1994 in Bangkok as a venue for multilateral dialogue on security issues in the Asia-Pacific region, is the institutional expression of the commitment to cooperative security. The ARF is an annual meeting (and associated processes) of 22 foreign ministers. The membership includes, in addition to the ASEAN countries, all major Asia-Pacific powers, including the United States, China, Japan, Russia, Korea, and Australia. Notwithstanding the ARF's modest agenda and slow pace, there is no support elsewhere in the region for an alternative form of multilateral security cooperation. For better or worse, therefore, the ARF appears to be a permanent fixture on the Asian-Pacific landscape. In this context, what are the prospects that the ARF might

---

[1]Paul Evans, "The Prospects for Multilateral Security Co-operation in the Asia/Pacific Region," *The Journal of Strategic Studies*, Vol. 18, No. 3, September 1995, pp. 201–217.

[2]Michael Leifer, "Truth About the Balance of Power," *Structure*, Singapore: Institute of Southeast Asia Studies, 1996, pp. 50–51.

draw China into cooperative relations with its neighbors and contribute to regional strategic stability?

In considering the ARF's potential for establishing an effective multi-lateral security regime, it is important to bear in mind that ASEAN launched the ARF initiative. It is not surprising, therefore, that these countries consciously styled the ARF's institutional machinery after ASEAN's goals, norms, procedures, and experiences. To the extent that ASEAN suffers from intrinsic defects as an effective subregional security organization, these limitations are mirrored in the ARF's cooperative security arrangements.[3]

Although observers of the ARF tend to judge its success or failure in terms of whether it prevents conflicts or solves regional security problems, the founders of the ARF set far more modest objectives for their novel enterprise, perhaps reflecting their understanding of the formidable obstacles to true multilateral security cooperation. Unlike European security institutions such as the Conference on Security and Cooperation in Europe (CSCE) and its successor organization, the Organization for Security and Cooperation in Europe (OSCE), the ARF is neither a negotiation process nor a collective security organization. Its objective is to improve the climate in which regional relations take place in an effort to manage bilateral and multilateral problems more effectively.[4] This conception of the ARF's purpose and utility is clearly reflected in the norms and functions established in the charter for the fledgling organization:

- The ARF was intended to evolve in three broad stages: the promotion of confidence-building, development of preventive diplomacy, and elaboration of approaches to conflict resolution.

- The ARF's rules of procedures would be based on "ASEAN norms and practices." Decisions would be made after careful and extensive consultations by consensus, without voting. In addition, the evolutionary approach of the ARF would progress "at a pace comfortable to all participants."

---

[3]Leifer, 1996, pp. 115–136.
[4]Lim.

- The ARF would play an essentially consultative security role, defined in comprehensive rather than narrow military terms, and would not try to impose solutions on its participants.

Within these parameters, therefore, the ARF's multilateral dialogue on security issues has evolved in accordance with the political traditions and cultural characteristics of the "ASEAN way." This distinctive approach to regional security emphasizes the process of building trust and confidence through consultation, dialogue, and transparency. Multilateral security cooperation is to be achieved incrementally, gradually, and informally, with primary emphasis on personal and political relationships rather than institutionalized arrangements. The ASEAN approach focuses on areas of common interest. Decisions are taken only by consensus and divisive issues are deferred for later resolution. This modus operandi has led one prominent Southeast Asian official to observe that "for the ARF, the process is just as important as any eventual agreement."[5]

ASEAN sensitivities to Chinese views regarding multilateral security cooperation have figured prominently in both the formulation of the ARF's guiding principles and the evolution of the ARF's practices and activities. Historically, participation in multilateral security arrangements has been anathema to Chinese leaders. This attitude reflects a realist view that China should rely on a balance of power to protect its security and avoid compromising Chinese freedom of action and sovereignty through entanglements in multilateral organizations.[6] In the Southeast Asian context, China prefers to deal with security issues and problems on a bilateral basis, in which Beijing would retain a significant advantage vis-à-vis any individual ASEAN country.[7] To ensure China's regular participation in ARF proceed-

---

[5]Almonte, pp. 81–82.

[6]Evans, p. 211.

[7]China has historically been cool to multilateral regional security arrangements. There have been recent indications, however, that Beijing has softened its opposition and is developing a "new security concept" for multilateral security cooperation that would gradually supplant the current regional security architecture, which is based on the U.S. bilateral alliances and American forward deployments. For details on the evolution of Chinese thinking on multilateral security arrangements for the Asia-Pacific region, see Banning Garrett and Bonnie Glaser, "Does China Want the U.S. Out of Asia?" *CSIS Pacific Forum PacNet Newsletter*, No. 22, May 30, 1997.

ings, the ASEAN countries have insisted that the ARF follow ASEAN's format and procedures—which gives China a significant lever to slow the pace of ARF activities—and acceded to Chinese requests to refrain from institutionalizing ARF activities.[8]

The ASEAN approach to regional security accounts in large measure for the difficulty it has experienced in forming a common approach to resolving the Spratly Islands problem. Most of the differences revolve around the priority that should be given to the form, structure, and content of discussions on cooperation in resolving the Spratlys claims, and the role of external powers and international organizations in mediating the dispute. Not surprisingly, in light of divergent ASEAN views and China's coolness toward multilateral or formal discussions on the Spratlys and internationalization of the issue, ASEAN's approach has been weighted toward the path of least resistance.[9]

Thus, the ASEAN strategy for conflict management in the South China Sea has revolved around participation since 1992 in Indonesian-sponsored informal, "unofficial" discussions of confidence-building measures (CBMs) and acceptance by the disputants of principles for resolving the conflict. The ASEAN approach appeared to produce results at the ARF meeting in Brunei in July 1995, where Chinese Foreign Minister Quian Quichen expressed China's readiness to discuss the issue with ASEAN, thereby reversing China's previous insistence that it would discuss it only bilaterally with individual states. Recent reports indicate, however, that China has gone back on its 1995 promise to discuss the dispute with all ASEAN claimants and has reverted to its original stand.[10]

The 1992 "ASEAN Declaration on the South China Sea" commits the signatories to use peaceful means to settle their disputes and to promote cooperation in joint development without prejudice to territorial claims. Nonetheless, this statement has had little practical effect.

---

[8]*Strategic Survey 1995–96*; "The Slow Progress of Multilateralism in Asia," *International Institute for Strategic Studies*, p. 191.

[9]Lee Lai To, "ASEAN and the South China Sea Conflicts," *The Pacific Review*, Vol. 8, No. 3, 1995, pp. 531–543.

[10]B. Raman, *Chinese Assertion of Territorial Claims. The Mischief Reef: A Case Study*, South Asia Analysis Group Papers, January 1999.

The renunciation of force does not go beyond what ASEAN states agreed to in the 1976 Treaty of Amity and Cooperation. China proposed joint development of offshore oil resources, but Beijing violated the spirit if not the letter of this formula by granting a drilling concession unilaterally to an American oil company in 1992 in waters claimed by Vietnam. Moreover, even though China also renounced the use of force in settling the Spratlys dispute, the Chinese interpretation of this commitment did not preclude their seizure of the unoccupied Mischief Reef in 1995 or subsequent unilateral Chinese moves to improve military installations there.

Thus, even though the current process would at best take years to achieve solid results, if they can be achieved at all, most ASEAN states will be reluctant to abandon this forum until China shows clear signs that it is not prepared to cooperate. That said, ASEAN states have been pragmatic and flexible in giving concrete meaning to the ASEAN approach to conflict management and prevention, as evidenced by the defense and military links they have formed with outside powers and the rather slow pace at which the ASEAN states have attempted to build a Zone of Peace, Freedom, and Neutrality (ZOPFAN) in the region.[11]

The ARF—or some other regional security architecture—may yet emerge as an effective mechanism for establishing regional order. However, the prospects for such a transformation in the immediate future are problematic at best. ASEAN's failure to play a role of any kind in the region's most serious crisis in recent years—the East Timor crisis of August/September 1999—induced many Southeast Asian security analysts to question the value of the organization as a regional security institution. For at least the next decade, the strategic environment in the Asia-Pacific region will be shaped primarily by the strategic interests of the major powers—the United States, China, and Japan. Barring a major shift in geopolitical relationships and the collapse of American political will and leadership, preserving

---

[11]China has been adept at stirring ASEAN hopes that multilateral dialogue on regional security issues, especially the dispute over the Spratlys, will bear fruit. Beijing, for example, has co-chaired ARF working group meetings with the Philippines on military CBMs and in 1996 the countries concluded an agreement on notification of military exercises. The publication in 1998 of the new Chinese Defense White Paper also represented another, albeit modest, step forward in transparency for the secretive PLA.

a strategic equilibrium in Southeast Asia will depend on U.S. bilateral security commitments and the maintenance of a balance of power based on deterrence and U.S. military might, and on the ability and willingness of the United States to contribute to preserving regional order and stability.

Chapter Seven

THE BALANCING ROLE OF THE UNITED STATES AND THE TAIWAN QUESTION

The ASEAN states rely in varying degrees on the U.S. military presence as a deterrent to Chinese military adventurism, but many in the area doubt the reliability of the United States as a security partner. As Southeast Asians often note, the United States could leave the region at any time, but they will always have to live with China. There are other political and diplomatic considerations that color the attitude of the ASEAN states toward military cooperation with the United States. In the case of several countries, notably Indonesia, the Philippines, and Malaysia, nationalist sentiment and lingering anticolonialist feelings have complicated the atmosphere for closer military relations with the United States. In addition, Indonesia and Malaysia fear that more visible military cooperation with the United States would tarnish their standing in the Non-Aligned Movement, and both are uneasy with the political burdens that accompany closer security ties with the United States. More fundamentally, however, many of the ASEAN governments, with the notable exception of Singapore and possibly Vietnam, worry that overly close military association with the United States would not only antagonize China but inflame domestic public opinion and exacerbate internal threats to political stability.[1]

Additionally, successful cooperation would have to overcome the gap in the perception of the proper roles and missions of the military in

---

[1] See Sheldon W. Simon, "East Asian Security: The Playing Field Has Changed," *Asian Survey*, Vol. 34, No. 12, December 1994, pp. 1047–1063.

the United States and Southeast Asia. In most ASEAN states, the military has a much broader sphere of responsibilities than in the United States and Western Europe. Since independence, for instance, the Indonesian armed forces have had an institutionalized role in governance—the sociopolitical function—which is only now beginning to change with the post-Suharto transformation of civil-military relations and the publication of a new political and military doctrine. In Thailand, the military for a long time played the role of political arbiter within a formally democratic political system. Cooperation with militaries that play a major political role raises difficult policy questions for the United States and other countries with a Western tradition of civil-military relations.

One important question in considering the future of U.S. military cooperation with the ASEAN states is whether any of those states would be prepared to increase peacetime military cooperation with the United States in response to Chinese military efforts to intimidate Taiwan. A related question is whether any Southeast Asian state would have a role to play in supporting U.S. defense of Taiwan against a Chinese attack or assisting the United States in patrolling and safeguarding the shipping lanes in a Taiwan conflict scenario.

A key factor will be the circumstances that trigger a PRC attack on Taiwan. If the immediate cause of the conflict is a Taiwanese declaration of independence, most if not all ASEAN countries will be reluctant to support Taiwan. If the attack is not perceived as provoked by Taiwan, however, the use of Chinese military force to intimidate Taiwan is likely to increase ASEAN fears of China and may make them more amenable to increased military cooperation with the United States. As a noted Southeast Asia security expert commented, China's bluster is tolerated by ASEAN as a sort of Chinese opera with much banging of gongs, but if a military blow is actually struck, ASEAN states will look to Washington while reevaluating their policies.[2] That said, it is unlikely that they would risk getting directly involved in a conflict with China by providing direct military assistance to Taiwan.

---

[2]Dr. Karl Jackson's comments to authors, February 2000.

Primarily for geopolitical reasons—e.g., the desire to maintain the United States as a counterweight to China—Singapore is likely to honor its bilateral agreements with the United States and provide logistical support to U.S. naval forces transiting from the Indian Ocean to the Taiwan Strait. It is unlikely, however, that Singapore or other ASEAN states would permit the United States to stage combat or combat support operations from their territory. Their unwillingness for direct involvement would stem from several factors. First, none would wish to make their country a target of possible Chinese military retaliation. Second, they would fear the long-term political, diplomatic, and economic repercussions of actions that China would consider an act of war. Third, many would face domestic opposition, especially from business interests, to allowing the United States to pull their country into what many perceive as an internal Chinese matter. Finally, because of geographic constraints, most of the ASEAN states would likely judge that their military support would have little impact on the outcome of a Taiwan conflict, while buying trouble with Beijing and at home.

The one possible exception to this outlook is the Philippines. Manila is likely to have fewer reservations than its ASEAN partners about tilting toward Taiwan, mainly because of its confrontation with China over the Spratly Islands and the lack of extensive economic and trade relations with China. Nonetheless, there would be substantial political costs to Manila in granting the United States use of its facilities to oppose Chinese military actions against Taiwan.

Therefore, in any dialogue on renewed U.S. access to Philippine bases for use in a Taiwan contingency, Manila would probably seek compensation for the risks associated with such an agreement. Compensation could include requests for security assistance or overt U.S. backing of the Philippines in its own territorial disputes with China. From a U.S. standpoint, these commitments would raise potentially serious problems, including budgetary constraints on new foreign aid commitments and congressional opposition to extending the scope of the U.S. security commitment in the South China Sea. In addition, the Philippine government would have to deal with negative domestic and regional reactions as well as China's wrath. In light of these pitfalls, while U.S. use of Philippine facilities in a Taiwan contingency cannot be ruled out, securing the use of these facilities would require the expenditure of substantial diplomatic and political capital.

To say that the ASEAN states are likely to react with caution to a U.S.-Chinese conflict over Taiwan is not to argue that they would be indifferent to how the United States would respond. Indeed, for all the ASEAN states the U.S. willingness to use force to defend Taiwan would be a test of the credibility of U.S. security commitments and U.S. ability to maintain the balance of power in the Asia-Pacific region. The other side of that coin is that, as a Singaporean political analyst noted, the United States would be right to question the value of its defense arrangements with ASEAN countries if it were denied the use of military facilities for the defense of vital U.S. security interests.[3]

---

[3] Discussion with panel of defense experts at the Institute of Defence and Strategic Studies, Singapore, February 2000.

Chapter Eight

IMPLICATIONS FOR U.S. STRATEGY AND DEFENSE PLANNING

For the foreseeable future, it is unlikely that the territorial and maritime disputes between China and some ASEAN states will be resolved through the multilateral conflict prevention and confidence-building measures proposed by the ASEAN Regional Forum. Thus, competing claims will pose a continuing risk of military conflict in the South China Sea. The political and social disruption brought about by the economic crisis has also increased the risk of disorder, piracy, and other transnational problems. Separatist movements threaten to destabilize Indonesia's fragile political transition and perhaps unleash a process of fragmentation. All of this creates opportunities for Beijing, if it were so inclined, to expand its presence and influence in the region. Further, over the next decade China will likely increase its current military advantage over any conceivable combination of ASEAN countries.

Nonetheless, several factors discussed in this study—China's economic priorities and dependence on foreign trade and investment, need for a stable regional environment, military shortcomings, and the possible impact on the Taiwan issue and on U.S. and Japanese defense policies—lessen the probability that China will use military force to achieve its political goals in Southeast Asia. Together, these factors suggest that the Chinese may not feel any particular sense of urgency to settle the South China Sea issue on Chinese terms. Indeed, the Chinese may well judge that time is on China's side, and that maintaining their control over some of the Spratly Islands, coupled with China's growing military and economic strength, will enable China to continue to probe for soft spots and make gains at little or negligible cost.

At the same time, as noted before, the possibility cannot be ruled out that conflict in the Spratlys could arise not from an act of premeditated aggression but rather from miscalculation or inadvertent escalation. Additionally, it is possible that China's growing economic integration into the global economy may not constrain aggressive Chinese behavior. For example, a beleaguered Chinese government, driven by domestic problems and rising Chinese nationalism, may continue its policy of "creeping assertiveness" or try to rally popular support by escalating a dispute with a neighboring state into a military conflict. Finally, the Chinese leadership might see an opportunity to dislodge adversaries from islands and seas it regards as its own and restore the Middle Kingdom's hegemonic position in this area.[1] What would be the consequences of Chinese success?

Exclusive Chinese possession of the Spratly Islands would threaten U.S. security interests if Beijing could use such control to: (1) deny the United States and other countries unrestricted access to the sea-lanes, or (2) impose Chinese domination over the region.

In peacetime, as already noted, the Chinese have no economic incentives to disrupt the sea-lanes. Further, even if China's intentions changed, the Chinese will not possess the military capabilities, at least for the next 10 to 15 years, to dominate the South China Sea in the face of determined U.S. military opposition.[2] Thus, a Chinese military threat to the sea-lanes seems plausible only as part of a broader conflict, such as in a PRC move against Taiwan.

Chinese possession of the Spratlys, while undesirable from a strictly military standpoint, is unlikely fundamentally to alter the military balance of power in the region or enhance China's ability to domi-

---

[1]See Shee Poon Kim, "The South China Sea in China's Strategic Thinking," *Contemporary Southeast Asia*, Vol. 19, No. 4, March 1998.

[2]Chinese understanding of these military and economic realities is reflected in official Chinese statements that forswear any intention of interfering with freedom of navigation in international waters and underscore that this position would not change even if Chinese claims in the South China Sea were validated.

nate the area.[3] Most of the islands cannot sustain the infrastructure for large-scale air and naval operations. Although it might be technically feasible to construct air and naval bases on a handful of islands, such facilities would be expensive to build and difficult to defend, given their vulnerability to air and missile attack or naval blockade. Chinese construction of military installations on the Spratlys would also be interpreted by many ASEAN states as provocative and a sign of hostile Chinese intent. Given this reaction, many of the Southeast Asian states would probably request U.S. military assistance. A positive U.S. response to these requests would cancel any Chinese military gains from turning the Spratlys into a forward base of operations.

Given the limited nature of the Chinese military threat to South China sea-lanes and chokepoints, the central issue for the United States and the ASEAN states is whether Chinese control of the Spratlys would be both a necessary and sufficient condition for establishing Chinese hegemony over the region. Chinese air bases closer to ASEAN countries would have military and perhaps coercive value, primarily by improving China's capability to provide air cover for Chinese naval forces operating in the South China Sea. That said, China is already significantly stronger than the ASEAN countries and Chinese control over the Spratlys would not appreciably change this equation. Likewise, whether China becomes a peer competitor of the United States in the next 15 to 20 years in Southeast Asia will be determined by a host of factors that have little to do with Chinese control of the Spratlys. Although energy reserves in the South China Sea could alleviate China's potential energy shortfall, those reserves constitute a tiny fraction of total global reserves. There is little danger, therefore, that China could use control of these energy resources to threaten global energy security for other coercive purposes.

Thus, there is little reason to fear that Chinese control of the Spratly Islands would give China the additional military, economic, or political muscle to achieve regional hegemony. Nonetheless, China's behavior in the South China Sea cannot be separated from broader Chinese strategic thinking. A Chinese decision to use force in the

---

[3]Robert Ross, "Beijing as a Conservative Power," *Foreign Affairs*, Vol. 76, No. 2, March/April 1997, pp. 36–37.

Spratlys could have symbolic importance as a barometer of Chinese intentions in the region. Even if Chinese control of the Spratlys fell short of giving China a platform to dominate the region, the successful use of force to attain political or military objectives would damage a major principle of international behavior—the non-use of force to settle international disputes—which has been at the heart of discussions between China and ASEAN countries over the South China Sea. It would also challenge the U.S. role of regional guarantor. The U.S. response to Chinese aggression in the Spratlys, regardless of the direct military consequences or America's intrinsic interests there, could be interpreted by both China and the ASEAN states as a signal of the U.S. willingness to use force to resist a broader Chinese geopolitical thrust in the region.

Consequently, to demonstrate to China that there is a cost to aggression and to discourage similar Chinese muscle flexing elsewhere in Asia—for instance, in Taiwan—the United States might want to take action to oppose Chinese military adventurism in the Spratlys and to reinforce the American commitment to regional stability and security. These reactions could be primarily diplomatic and economic (e.g., sanctions, statements), but there could also be a military dimension, including additional military deployments to the region,[4] arms transfers, and increased military contacts.[5]

Of course, whether or not China gains control of the Spratlys, it could emerge as an aggressive hegemonic threat in the future. Should this occur, the countries of Southeast Asia would face a serious dilemma. For the foreseeable future, the ASEAN states will confront a huge gap between their external security needs and their capabilities to meet these needs, particularly if a prolonged economic downturn leads to more aggressive Chinese behavior and a deterioration of ASEAN cohesion and military capabilities. As one well-known expert on the

---

[4]Five months after the Mischief Reef incident, a contingent of U.S. Navy Seals arrived in Puerto Princesa, headquarters of the Philippines Western Military Command, to train Filipino troops stationed in areas in the Spratlys under Filipino control. According to some analysts, the prospects of revived U.S.-Filipino military cooperation appeared to have had a sobering effect on Beijing. Raman, pp. 9–10.

[5]Some analysts have also suggested that U.S. air and naval forces could assist ASEAN countries in military operations to evict Chinese forces, perhaps by supporting a blockade or conducting air strikes on military targets on the islands. David A. Shlapak and David T. Orletsky, "China's Military in Transition," internal 1998 RAND paper.

region has observed, none of the three options available to bolster external security offers great promise.[6]

- **National self-defense.** With the possible exception of Indonesia, individual ASEAN states simply lack the size, resources, and capabilities to sustain a military buildup that could offset Chinese military preponderance. For the foreseeable future, Indonesia has been too weakened by the economic crisis and political instability to generate a credible deterrent capability.

- **Collective self-defense.** There is deep and abiding opposition to multilateral military cooperation within ASEAN or formation of an ASEAN defense pact. Although greater multilateral security cooperation may evolve to deal with low-level threats (e.g., piracy, smuggling, protection of EEZs), ASEAN now lacks the cohesion and capabilities to counter serious military threats.[7]

- **Regional security arrangements.** Given the constraints on individual and collective self-defense, ASEAN states have searched for broader regional security arrangements that might deter China or resolve disputes that might lead to armed conflict. Thus far, however, the creation of a viable and effective regional security structure has proved elusive. Even if ASEAN could overcome the many intra-ASEAN obstacles to such a regional order, China's interest in effective multilateral security arrangements that are not weighted in Beijing's favor remains a question mark.[8]

In other words, if an aggressive and hostile China sought to achieve regional hegemony the ASEAN states are likely to have no viable alternative to reliance on U.S. military forces to deter aggressive Chinese behavior, unless they decide to throw their lot in, or

---

[6]Leszek Buszynski, "ASEAN Security Dilemmas," *Survival*, Vol. 39, No. 4, Winter 1992–1993, pp. 90–107.

[7]On the other hand, Dr. Karl Jackson noted that ASEAN functioned as a multilateral security community throughout the struggle in Cambodia. Comments to authors, February 2000.

[8]For an examination of Chinese attitudes toward multilateral security cooperation, see Banning Garrett and Bonnie Glaser, "Multilateral Security in the Asia-Pacific Region and Its Impact on Chinese Interests: Views from Beijing," *Contemporary Southeast Asia*, Vol. 16, No. 1, June 1994, pp. 14–34. See also fn. 8 of Chapter One.

"bandwagon," with China. Given the uncertainty about China's future strategic direction, therefore, the issue is not whether the United States should seek to establish a prudent hedge in Southeast Asia against the possibility of an adversarial China. Rather, the key issues revolve around managing the implementation of this hedge strategy—the timing, content, and sequencing of hedging actions, the relationship of these measures to broader policies of engagement and containment, the resources that should be expended to establish a hedge, and the risks associated with moving too slowly or rapidly in taking hedging actions.

Without a fundamental change in threat perceptions of China and a resolution of internal problems and intra-ASEAN frictions, ASEAN will likely persevere with its current approach of dialogue, cooperation, engagement, and expanded economic interdependence to restrain Chinese ambitions. At the same time, however, some ASEAN states may seek reassurance from the United States and tangible signs of U.S. military support. In the short run, such requests are likely to be modest and intended primarily to "keep China honest" rather than create a robust war-fighting capability through the establishment of U.S. bases or a permanent land-based U.S. military presence.[9]

Nonetheless, any increase in U.S. peacetime military activities in Southeast Asia could make an important difference on the margins of Chinese strategic calculations, primarily because of China's continuing military weaknesses vis-à-vis the United States. While U.S. naval forces will play the primary role in a South China Sea contingency, access to the region for U.S. land-based fighter aircraft would complicate Chinese calculations, because of the serious difficulties China would face in establishing air superiority for its naval forces operating in the South China Sea.

These considerations have several implications for U.S. defense planning and the USAF:

---

[9]Tangible signs of U.S. support could include, for instance, U.S. military reengagement with the Philippines now that the Visiting Forces Agreement has been ratified by the Philippine Senate; willingness to transfer NATO-releasable advanced military technology to states with which the United States has a close and ongoing military relationship; and cooperation with ASEAN states on counterterrorism and regional order-keeping initiatives.

- First, the United States should think in terms of a step-by-step approach to hedging. The initial phase of a hedging strategy should focus on shaping a more favorable security environment through engagement, dialogue, reassurance, and trust-building.

- Second, over the next several years the United States will have an opportunity to cultivate stronger military ties with many ASEAN states and perhaps to play a behind-the-scenes role in facilitating closer intra-ASEAN defense cooperation. Military-to-military contacts should put priority on encouraging professionalism and modernization in a democratic context. Indonesia's democratic evolution since the fall of Suharto has opened a window of opportunity for closer military-to-military ties with the Indonesian armed forces (TNI), and the scope of bilateral military cooperation could widen in a post-Mahathir Malaysia. The priority during this time frame should be to increase military engagement to foster habits of cooperation and interoperability. China might even be included in some of these activities as a transparency and confidence-building measure.

- Third, until the Southeast Asian economies emerge from the economic crisis, the United States should restore a robust security assistance program to allies in the region, particularly the Philippines. Providing urgently needed air defense and naval patrol assets to the Philippines would help Manila to reestablish deterrence vis-à-vis China and give a further impetus to the revitalization of the United States-Philippine defense relationship. The United States should also restore full military-to-military ties with Indonesia and resume the transfer of military equipment and spare parts needed to prevent the further deterioration of Indonesian defense capabilities.

- Fourth, there are a number of low-key but valuable steps that the USAF could consider to expand military cooperation, trust, and confidence with ASEAN militaries. One especially fruitful approach would be to expand military-to-military contacts and training to assist ASEAN countries with the modernization of their air forces and the use of the assets to combat illicit drug trafficking, smuggling, and piracy. The U.S. program of engagement with Singapore could serve as a model to expand pilot

training and officer exchanges. Exercise *Cope Thunder* could
also be expanded to include other ASEAN countries.[10] The USAF
could increase periodic deployments of airborne warning and
control system (AWACS) E-3 Sentry aircraft for training in a
maritime surveillance mode with ASEAN military units.
Additionally, the USAF could begin a dialogue on bilateral and
regional cooperation to improve the effectiveness of anti–drug-
smuggling operations, the delivery of disaster relief, and re-
sponses to environmental disasters. Specifically, these talks
could address U.S. technical assistance in establishing a regional
air surveillance network. All of these contacts would offer substantial mutual benefits without threatening China. Indeed,
China could be invited to participate in some of these activities.
At the same time, these interactions would help establish an improved atmosphere for closer United States-ASEAN military cooperation if warranted by the nature and direction of Chinese
policies.

- Finally, given the near-term political constraints on significant
  ASEAN military cooperation with the United States, military and
  diplomatic planners should adopt a "portfolio approach" toward
  access and basing arrangements. In other words, as long as there
  is clear risk that internal instabilities and weak ASEAN governments could threaten loss of, or timely and unhindered access to,
  facilities, the United States should seek as much diversification
  as possible in its regional military infrastructure, consistent with
  operational and budgetary considerations.

For the next 5 to 10 years, assuming a continuation of current trends,
it should be possible to expand military-to-military contacts in
meaningful ways. These activities could include cooperation in sea-
monitoring, search and rescue, and combined exercises. Ideally, the
USAF and U.S. Navy would exercise and train together in the region
with more than one ASEAN country and perhaps Australia and the
UK under the aegis of the Five Power Defense Arrangement, if this
agreement survives current tensions between Singapore and

---

[10]*Cope Thunder* was conducted by Pacific Air Forces (PACAF) in the Philippines prior
to the U.S. withdrawal from bases there, and subsequently in Alaska. Participants in
the most recent exercises include Japan, the Philippines, Australia, New Zealand,
Thailand, and Singapore.

Malaysia. Singapore would be an attractive candidate because of its location, military professionalism, and technical sophistication. However, the United States should seek to involve some other ASEAN country because Singapore's space limitations could constrain air operations. The Philippines, Thailand, Malaysia, and/or Indonesia are possibilities for such expanded multilateral training and exercises.

A major challenge in setting priorities will be reconciling political constraints on access/basing with USAF operational requirements in specific contingencies. Political considerations argue for spreading access/basing arrangements among several countries to avoid overdependence on any single country and to hedge against the possible loss of access or onerous operational restrictions placed on U.S. forces arising from political sensitivities in host countries. A key question is whether this diversification strategy is compatible with a viable operational concept for supporting expeditionary operations if the latter required a greater concentration of assets and infrastructure.

The sea-lanes through Southeast Asia are vulnerable in two areas: the Straits of Malacca, Sunda, and Lombok and the two main shipping channels in the South China Sea running east and west of the Spratly Islands. The ideal bases of operation to gain control of the air space over the Straits are Singapore, Malaysia, and Indonesia. The Philippines and Vietnam would be suitable for support of USAF operations to establish air superiority over the main shipping channels in the South China Sea. As suggested earlier, the U.S. Navy would play the predominant role in defense of the sea-lanes in Southeast Asia, especially the main shipping channels through the South China Sea. However, the USAF could play a critical role, particularly to the degree that U.S. carrier battle groups (CVBGs) might have difficulty operating in the confined spaces of the Straits and a large portion of U.S. naval forces might be preoccupied with countering the Chinese submarine, mining, and surface naval threat to the sea-lanes.

From a strictly operational standpoint, therefore, the U.S./USAF priority, in conjunction with the U.S. Navy, should be to improve military ties and cooperation with the Philippines, Singapore, Malaysia, Indonesia, and Vietnam. Because there is no urgency to establishing U.S. military bases, and the ASEAN states are ultimately dependent on the United States to maintain a balance of power in the region, the United States need not act as the *demandeur* in trying to forge stronger military relations with these countries. Particularly with Malaysia and Indonesia—both of which are committed to national and regional "self-reliance" and are sensitive about their sovereignty and position within the Non-Aligned Movement—the United States/USAF will need to be patient in building trust in the relationship and in improving their defense capabilities. Enhanced U.S. intelligence sharing and arms transfers, especially those that improve interoperability with U.S. forces, as well as U.S. assistance tied to improving intra-ASEAN military cooperation, could pave the way for expanded military cooperation should threat perceptions of China change. Furthermore, by pursuing a diversification strategy the United States should be able to maximize its bargaining leverage with each country.

From a political standpoint, however, the United States might encounter serious constraints in raising its military profile with Indonesia and Malaysia. Indeed, should China emerge as an aggressive and expansionist threat in the next 10 to 15 years, Singapore, the Philippines, and possibly Vietnam may prove more amenable to hosting an increased U.S. military presence, depending of course on the overall state of U.S. bilateral relations with these countries and regional political dynamics. Singapore is ideally located for protection of the Straits, but the distance of the Philippines and Vietnam from the strategic chokepoints of Southeast Asia would reduce their operational value in any effort to prevent closure of the Straits. However, access to both countries (e.g., for staging/bed-down of combat aircraft or tanker/AWACS support) would help to establish air superiority over the sea-lanes of the South China Sea. Moreover, any Chinese air and naval assets engaged in a campaign to close the Straits would need to transit areas of the South China Sea that would be vulnerable to air interdiction. For these reasons, therefore, the U.S. hedging strategy in Southeast Asia should seek to encompass the Philippines and Vietnam as well as Indonesia, Malaysia, and Singapore.

A potential political constraint on a U.S. engagement strategy with the ASEAN militaries will be the overall level of democracy and human rights practices in the respective ASEAN countries. The military's involvement in political and internal security activities in some

ASEAN countries, particularly Indonesia during the Suharto era, created barriers to military-to-military cooperation with the United States. At the same time, the militaries in most ASEAN countries are important, and sometimes dominant, players in the political system, as well as in defense and security policy decisions. The United States therefore needs to walk a fine line between the need to engage ASEAN militaries and influence their values, security doctrines, and political actions and to avoid association with questionable activities.

The residual effects of the East Timor crisis, the insurgencies in Aceh and West Papua (Irian Jaya), and ethnic and religious strife in the Moluccas and elsewhere have the potential to derail the fragile political transition in Indonesia, as well as the prospects for cooperation with the United States. An unstable Indonesia would not make a suitable security partner for the United States. Moreover, the resulting geopolitical vacuum could draw in external powers such as China and increase the demands on the USAF.

In conclusion, without clear and unambiguous indications that China seeks to overturn the status quo, many ASEAN states will be reluctant to arouse Chinese antagonism by taking actions that China would regard as provocative; in general, therefore, Southeast Asian states will prefer to restrain China through a combination of economic integration and diplomatic engagement. On the other hand, during this time frame, fears of a rising China could lead some ASEAN states to seek reassurance of the U.S. commitment to regional security, including an expanded U.S. military presence in the region.

Given these complex and somewhat contradictory attitudes, how should the United States treat Southeast Asia in the context of its evolving strategy toward China? In light of the uncertainty surrounding China's future strategic orientation, what is the appropriate balance between the broader policy of engagement with China and hedging activities with individual ASEAN countries? How far should the United States go in hedging against the possible emergence of a hostile China? What are the risks of pursuing hedging activities at a deliberate pace?

At least for the next 5 to 10 years, China will pose a limited military threat to U.S. security interests in Southeast Asia, for the reasons described in this study. Because the Chinese threat will evolve gradually, U.S. military cooperation with the ASEAN states and the expansion of the U.S. military presence can proceed at a deliberate pace. The primary purpose of U.S. peacetime military activities should be shaping; in other words, the aim of any increase in military contacts and arrangements with host countries should be to create a more secure strategic environment rather than the infrastructure to support U.S. war-fighting capabilities in a military engagement with China. Should China abandon moderation in favor of a hostile course toward its neighbors, U.S. "shaping" activities can establish a better climate for robust military cooperation with ASEAN states aimed at thwarting expansive Chinese geopolitical ambitions—in short, a "virtual alliance."[11] The country priorities for these shaping and hedging activities should be the Philippines and Singapore, followed by Malaysia, Indonesia, and Vietnam.

The gradualist approach described above promises low risks and a potentially high payoff. An overly aggressive and ambitious effort to secure access to facilities in support of a major expansion of the U.S. peacetime military presence and power projection capabilities would antagonize China, add an unnecessary irritant to U.S. bilateral relations with ASEAN states, and widen intra-ASEAN differences.

The potential downside to such a "go slow" approach is that the United States could be caught flat-footed if the Chinese military threat materializes faster than most observers anticipate. As other RAND studies have suggested, however, it is unlikely that China will achieve a "leap-ahead" breakthrough in military capabilities during the next decade. Moreover, a shaping and hedging approach offers two significant benefits: first, raising the U.S. military profile in the region, even if done in only modest ways, will reinforce Beijing's caution and underpin the credibility of our security commitments. Second, an incremental approach to improving U.S. military relationships with ASEAN states would avoid the pitfalls of a premature policy of "containing" China while capturing the benefits of hedging—specifically, laying the groundwork for implementing multi-

---

[11]See Saunders.

lateral security cooperation to cope with a Chinese threat to the security of Southeast Asia and to U.S. vital interests in the region should one emerge.

ILLUSTRATIVE ASIAN ECONOMIC SCENARIOS

1. THE WORST IS OVER

In this scenario, there is no second round of the Asian financial crisis. Growth returns to the region, but inevitably at a slower pace than in the 1980s and 1990s. Japan slowly cleans up its banking troubles and resumes the growth rates expected of a mature, industrialized economy. Nevertheless, with an aging population and underfunded retirement obligations, the Japanese government will find deficit spending increasingly difficult. Both defense spending and aid to other Asian countries will likely decrease. Korea, Taiwan, Hong Kong, and Southeast Asia slowly get their respective houses in order and resume moderate growth, but the "Asian economic miracle" is over. With little further ability to draw on underutilized factors of production, growth will slow to the 4 to 5 percent range—enough to generate noticeable improvements in standards of living, but not enough for governments to be confident that the rising tide is in fact lifting all boats. Governments will have to pay increasing attention to distribution questions, worrying for the first time about how to build effective social safety nets.

As the general Asian crisis recedes, China is able to avoid another devaluation. Exports will recover, but growth rates will fall under 8 percent. Chinese policymakers will slowly begin to close down state-owned enterprises and, as a result, will face a growing problem of displaced labor. Foreign investors will be less attracted to China than in the recent past, troubled by a still fragile banking system and a de-

gree of transparency in financial matters that lags considerably be-
hind improving international practice.

In this scenario, the attention of Asian policymakers will be focused
principally on domestic concerns. These policymakers will have little attention and fewer resources to spare for regional initiatives.
Trade liberalization will proceed slowly in the Asia Pacific Economic
Cooperation (APEC) framework. There will be occasional small-scale
crises, as one or another country hits bumps on the road to recovery.
Without prospects of a meaningful regional response, countries that
get into trouble will continue to look to international financial institutions for assistance when needed. U.S. economic policy in the region will focus on making progress on trade liberalization and encouraging Asian countries to stick with the "Washington consensus"
of free trade and capital movements, privatization, deregulation, and
fiscal and monetary restraint. But U.S. efforts will be mostly exhor-
tation and leading by example; little material assistance will be forth-
coming.

2. FURTHER DETERIORATION—CHINA BECOMES THE LATEST VICTIM OF THE ASIAN CRISIS

This scenario assumes further economic deterioration and is the
most unpredictable scenario with regard to the modalities it might
take. The scenario could come about if continued malaise in Japan
and the rest of Asia further depresses Chinese exports. Fears of a
Chinese devaluation encourage capital flight. Chinese growth falls
below the rates required to maintain employment in all parts of the
country. China devalues its currency, hoping that a one-time devaluation will be sufficient to restore export competitiveness and to discourage further capital flight.

Southeast and South Asian countries that compete with China are
hard hit. Fragile recoveries are aborted and some of the affected
countries respond with currency devaluations. Financial uncertainty
again sweeps the region. As in the 1930s, this has profound political
and regional security consequences. Within the region, there will be
an increase in civil unrest and some of the more fragile governments
will face threats to their survival. In official and commercial circles,
there will be considerable resentment of China for touching off an-

other round of competitive devaluations. Relations between China and other Asian nations will come to be dominated by bickering over trade matters, illegal immigration, and possibly the treatment of the overseas Chinese communities.

In China, a sharp decline in economic performance could threaten the viability of the country's banking system, sharpen regional differences, and possibly bring into question the legitimacy of the regime. The role of the People's Liberation Army in the polity will grow. There could be power shifts within the military, with significant consequences for foreign and security policy decisions.

In this scenario, with Korea and Japan mired in their own economic problems, the United States will be the only country capable of exercising leadership. Risks and opportunities for the United States will be high. Both U.S. influence and demands on U.S. capabilities—in the political, economic, and military spheres—will grow. U.S. influence will be limited only to the quality of U.S. leadership and the strength of the U.S. commitment to the region.

3. CHINA DODGES THE BULLET

China somehow manages to get its house in order. It begins to make progress in reforming state-owned enterprises; it gains real control over International Trade and Investment Corporations (ITICs); and it begins to reform its banking system. Japan, on the other hand, continues to muddle through without breaking out of its economic paralysis. Without the Japanese locomotive to pull the train, other Asian economies cannot mount sustained recoveries. Within Southeast Asia, there is greater differentiation in economic performance, depending on the individual countries' ability to maintain political stability and appropriate economic and fiscal policies.

A newly confident China begins to play a larger role in Asian economic affairs, mobilizing its large foreign currency reserves to assist other Asian countries whose currencies come under attack. There will be greater opportunities for the expansion of Chinese influence and lesser ability on the part of regional states to balance it.

The United States finds itself in an uneasy alignment with China in trying to stabilize the Asian economies. Beijing continues to point out that maintenance of the value of the yuan is not without cost to China. Chinese officials declare that China will stay the course for the good of the region, as long as other countries do not try to take unfair advantage of China's forbearance. China becomes more assertive as to what constitutes unfair advantage and the sort of policies it finds unacceptable. Increased consultation and even coordination with the United States and the international financial institutions would be helpful in this regard.

4. THE REST OF ASIA RECOVERS, BUT CHINA FALTERS

This scenario is the reverse of Scenario 3. Japan takes credible steps to revitalize its banking system and resume higher rates of economic growth. The Southeast Asian economies begin to recover, but China fails to deal with its structural problems in the banking and state industrial sectors. As in Scenario 2, an economic crisis in China would have significant and unpredictable consequences.

There are several paths that Beijing could take in these circumstances. It could:

- Turn inward and concentrate on resolving its internal difficulties—through processes of either liberalization or repression.

- Pursue foreign adventures in hopes of generating domestic unity. In this subscenario, Taiwan could be a flashpoint. The PRC could also become more assertive in pressing its claims to the South China Sea, or it could take up the defense of beleaguered ethnic Chinese communities in Southeast Asia.

- Or China may be unable to maintain domestic stability and it could enter a period of increased turmoil and even civil war. In this subscenario, Beijing would face increased separatist challenges in Xinjiang and Tibet. Taiwan might see this as a window of opportunity to assert its independence.

SELECTED BIBLIOGRAPHY

Acharya, Amitav, "Regional Military-Security Cooperation in the Third World: A Conceptual Analysis of the Relevance and Limitations of ASEAN," *Journal of Peace Research*, Vol. 29, No. 1, 1992, pp. 7–21.

_____, "A Regional Security Community in Southeast Asia?" *The Journal of Strategic Studies*, Vol. 18, No. 3, September 1995, pp. 175–200.

_____, "A Concert of Asia?" *Survival*, Vol. 41, No. 3, Autumn 1999, pp. 84–99.

Alagappa, Muthiah (ed.), *Asian Security Practice: Material and Ideational Influences*, Stanford, CA: Stanford University Press, 1998.

Almonte, Jose T., "Ensuring Security the ASEAN Way," *Survival*, Vol. 39, No. 4, Winter 1997–1998, pp. 80–92.

Asian Development Bank, *Asian Development Outlook 1999*, update, Manila, September 1999.

Ball, Desmond, "Strategic Culture in the Asia-Pacific Region," *Security Studies*, Vol. 3, No. 1, Autumn 1993, pp. 44–74.

Bellows, Michael D., *Asia in the 21st Century: Evolving Strategic Priorities*, Washington, D.C.: National Defense University Press, 1994.

Bernstein, Richard, and Ross H. Munro, "The Coming Conflict with America," *Foreign Affairs*, Vol. 76, No. 2, March/April 1997, pp. 18–32.

Betts, Richard K., "Vietnam's Strategic Predicament," *Survival*, Vol. 37, No. 3, Autumn 1995.

Breckon, Lyall, *The Security Environment in Southeast Asia and Australia, 1995–2010*, Alexandria, VA: Center for Naval Analyses, March 1996.

_____, and Thomas J. Hirschfield, *The Dynamics of Security in the Asia-Pacific Region*, Alexandria, VA: Center for Naval Analyses, January 1996.

Burles, Mark, and Abram N. Shulsky, *Patterns in China's Use of Force*, RAND, MR-1160-AF, 2000.

Buzan, Barry, and Gerald Segal, "Rethinking East Asian Security," *Survival*, Vol. 36, No. 2, Summer 1994, pp. 3–21.

Caldwell, John, *China's Conventional Military Capabilities, 1994–2004*, Washington, D.C.: The Center for Strategic and International Studies, 1994.

Carpenter, William M., and David G. Wiencek (eds.), *Asian Security Handbook: An Assessment of Political-Security Issues in the Asia-Pacific Region*, Armonk, NY: M. E. Sharpe, Inc., 1996.

_____, "Openness and Security Policy in South-east Asia," *Survival*, Vol. 38, No. 3, Autumn 1996, pp. 82–98.

Cheng, Felix K., "Beijing's Reach in the South China Sea," *Orbis*, Summer 1996, pp. 353–374.

Christensen, Thomas J., "Chinese Realpolitik," *Foreign Affairs*, Vol. 75, No. 5, September/October 1996, pp. 37–52.

Chu, Shulong, "The PRC Girds for Limited, High-Tech War," *Orbis*, Spring 1994, pp. 177–191.

Clad, James, "Fin de Siecle, Fin de l'ASEAN?" *CSIS PacNet Newsletter*, No. 9, March 3, 2000.

Clifford, Mark L., and Pete Engardio, *Meltdown: Asia's Boom, Bust, and Beyond*, Upper Saddle River, New Jersey: Prentice Hall, 2000.

Cossa, Ralph A., "Security Implications of Conflict in the South China Sea: Exploring Potential Triggers of Conflict," *CSIS PacNet Newsletter*, No. 16, April 17, 1998.

Cronin, Patrick, and Marvin Ott, *The Indonesian Armed Forces: Roles, Prospects, and Implications*, Strategic Forum Paper No. 126, Washington, D.C.: National Defense University, August 1997.

Da Cunha, Derek (ed.), *The Evolving Pacific Power Structure*, Singapore: Institute of Southeast Asian Studies, 1996.

_____, "Southeast Asian Perceptions of China's Future Security Role in Its 'Backyard,'" in Jonathan D. Pollack and Richard H. Yang (eds.), *In China's Shadow*, RAND, CF-137-CAPP, 1998, pp. 115–126.

DeWitt, David, and Brian Bow, "Proliferation Management in Southeast Asia," *Survival*, Vol. 38, No. 3, Autumn 1996, pp. 67–81.

Dibb, Paul, *Towards a New Balance of Power in Asia*, Adelphi Paper 295, Oxford: Oxford University Press for IISS, 1995.

_____, "The Revolution in Military Affairs and Asian Security," *Survival*, Vol. 39, No. 4, Winter 1997–1998, pp. 93–116.

Dillon, Dana R., "Contemporary Security Challenges in Southeast Asia," *Parameters*, Spring 1997, pp. 119–133.

Eikenberry, Karl W., "Does China Threaten Asia-Pacific Regional Stability?" *Parameters*, Spring 1995, pp. 82–103.

Ellings, Richard J., and Sheldon W. Simon (eds.), "Southeast Asian Security in the New Millennium," *The National Bureau of Asian Research*, 1996.

Evans, Paul M., "The Prospects for Multilateral Security Co-operation in the Asia/Pacific Region," *The Journal of Strategic Studies*, Vol. 18, No. 3, September 1995, pp. 201–217.

Everett, Michael W., and Mary A. Sommerville (eds.), *Multilateral Activities in Southeast Asia*, Washington, D.C.: National Defense University Press, 1995.

Far Eastern Economic Review, "Imperial Intrigue," Vol. 160, No. 37, September 11, 1997, pp. 14–15.

Feigenbaum, Evan A., "China's Military Posture and the New Geopolitics," *Survival*, Vol. 41, No. 2, Summer 1999.

Fisher, Richard D. Jr., "The Accelerating Modernization of China's Military," in Kim R. Holmes and James J. Przystup (eds.), *Between*

Diplomacy and Detente: Strategies for U.S. Relations with China,
Washington, D.C.: The Heritage Foundation, 1997, pp. 97–140.

_____, "Does China Want the U.S. Out of Asia?" *CSIS Pacific Forum PacNet Newsletter,* No. 22, May 30, 1997.

Godwin, Paul H.B., "Force Projection and China's National Military Strategy," in C. Dennison Lane, Mark Weisenbloom, and Dimon Liu (eds.), *Chinese Military Modernization,* New York: Kegan Paul International, 1996, pp. 69–99.

_____, "From Continent to Periphery: PLA Doctrine, Strategy, and Capabilities Towards 2000," *The China Quarterly,* No. 146, June 1996, pp. 464–487.

Goldstein, Avery, "Great Expectations: Interpreting China's Arrival," *International Security,* Vol. 22, No. 3, Winter 1997–1998, pp. 36–73.

Goodman, David S.G., "Are Asia's 'Ethnic Chinese' a Regional Security Threat?" *Survival,* Vol. 39, No. 4, Winter 1997–1998.

Gregor, A. James, "China, the United States, and Security Policy in East Asia," *Parameters,* Summer 1996, pp. 91–101.

Hull, Richard E., *The South China Sea: Future Source of Prosperity or Conflict in Southeast Asia?* Strategic Forum Paper No. 60, Washington, D.C.: National Defense University Press, February 1996.

Huxley, Tim, "Singapore and Malaysia: A Precarious Balance? *Pacific Review,* Vol. 4, No. 1, Autumn 1991.

_____, and Susan Willett, *Arming East Asia,* Adelphi Papers 329, Oxford: Oxford University Press, 1999.

International Monetary Fund, *Direction of Trade Yearbook,* 1999.

Jane's Defence Weekly, various issues.

Jane's International Defense Review, April 1997 and September 1997.

Jie, Chen, "China's Spratly Policy," *Asian Survey,* Vol. 34, No. 10, October 1994, pp. 893–903.

Kenny, Henry J., *An Analysis of Possible Threats to Shipping in Key Southeast Asian Sea Lanes,* Alexandria, VA: Center for Naval Analyses, 1996.

Khalilzad, Z., A. Shulsky, D. Byman, R. Cliff, D. Orletsky, D. Shlapak, and A. Tellis, *The United States and a Rising China: Strategic and Military Implications*, RAND, MR-1082-AF, 1999.

Kim, Samuel S., "China's Quest for Security in the Post-Cold War World," Carlisle Barracks, PA: Strategic Studies Institute, U.S. Army War College, July 1996.

Kim, Shee Poon, "The South China Sea in China's Strategic Thinking," *Contemporary Southeast Asia*, Vol. 19, No. 4, March 1998.

Lim, Robyn, "The ASEAN Regional Forum: Building on Sand," *Contemporary Southeast Asia*, Vol. 20, No. 2, August 1998, pp. 115–136.

Lin, Chong-pin, "The Power Projection Capabilities of the People's Liberation Army," in C. Dennison Lane, Mark Weisenbloom, and Dimon Liu (eds.), *Chinese Military Modernization*, pp. 100–125.

Mak, J. N., "The Modernization of the Malaysian Armed Forces," *Contemporary Southeast Asia*, Vol. 19, No. 1, June 1997.

The Military Balance 1998/99, International Institute for Strategic Studies, Oxford: Oxford University Press.

Munro, Ross H., "Eavesdropping on the Chinese Military: Where It Expects War—Where It Doesn't," *Orbis*, Summer 1994, pp. 355–372.

_____, "China's Waxing Spheres of Influence," *Orbis*, Fall 1994, pp. 586–605.

Narine, Shaun, "ASEAN and the ARF: The Limits of the 'ASEAN Way,'" *Asian Survey*, Vol. 37, No. 10, October 1997, pp. 961–978.

Noer, John H., *Southeast Asian Chokepoints*, Strategic Forum Paper 98, Washington, D.C.: National Defense University, December 1996.

_____, *Chokepoints: Maritime Economic Concerns in Southeast Asia*, Washington, D.C.: National Defense University, 1996.

Nye, Joseph S., "China's Re-emergence and the Future of the Asia-Pacific," *Survival*, Vol. 39, No. 4, Winter 1997–1998, pp. 65–79.

Pillsbury, Michael, "Dangerous Chinese Misperceptions: The Implications for DoD," unpublished paper prepared for the Office of Net Assessment, U.S. Department of Defense, 1997.

Pollack, Jonathan D., "Pacific Insecurity: Emerging Threats to Stability in East Asia," *Harvard International Review*, Vol. 15, No. 2, Spring 1996, pp. 8–11.

_____, "Designing a New American Security Strategy for Asia," in James Shinn (ed.), *Weaving the Net: Conditional Engagement with China*, New York: Council on Foreign Relations, 1996, pp. 99–132.

_____, and Richard H. Yang (eds.), *In China's Shadow: Regional Perspectives on Chinese Foreign Policy and Military Development*, RAND, CF-137-CAPP, 1998.

Raman, B., *Chinese Assertion of Territorial Claims. The Mischief Reef: A Case Study*, South Asia Analysis Group Papers, January 1999.

Republic of Singapore, *Defending Singapore in the 21st Century*, Ministry of Defence, 2000.

Ross, Robert S., "Beijing as a Conservative Power," *Foreign Affairs*, Vol. 76, No. 2, March/April 1997, pp. 33–44.

Roy, Denny, "Hegemon on the Horizon? China's Threat to East Asian Security," *International Security*, Vol. 19, No. 1, Summer 1994, pp. 149–168.

Salameh, Mamdouh G., "China, Oil, and the Risk of Regional Conflict," *Survival*, Vol. 37, No. 4, Winter 1995–1996, pp. 133–146.

Saunders, Phillip C., "A 'Virtual Alliance' for Asian Security," *Orbis*, Spring 1999.

Segal, Gerald, "East Asia and the 'Constrainment' of China," *International Security*, Vol. 20, No. 4, Spring 1996, pp. 107–135.

Shambaugh, David, "The United States and China: A New Cold War?" *Current History*, September 1995, pp. 241–247.

_____, "China's Military: Real or Paper Tiger?" *The Washington Quarterly*, Spring 1996, pp. 14–36.

_____, "Containment or Engagement of China?" *International Security*, Vol. 21, No. 2, Fall 1996, pp. 180-209.

Simon, Sheldon W., *The Economic Crisis and ASEAN States' Security*, Carlisle Barracks, PA: Strategic Studies Institute, U.S. Army War College, 1998.

Smith, Anthony, "Indonesia's Role in ASEAN: The End of Leadership?" *Contemporary Southeast Asia*, Vol. 21, No. 2, August 1999.

Smith, Esmond D., "China's Aspirations in the Spratly Islands," *Contemporary Southeast Asia*, Vol. 16, No. 3, December 1994, pp. 274–294.

Snitwongse, Kusuma, "ASEAN's Security Cooperation: Searching for a Regional Order," *The Pacific Review*, Vol. 8, No. 3, 1995, pp. 518–530.

Stares, Paul, and Nicolas Regaud, "Europe's Role in Asia-Pacific Security," *Survival*, Vol. 39, No. 4, Winter 1997–1998.

Storey, Ian James, "Creeping Assertiveness: China, the Philippines, and the South China Sea Dispute," *Contemporary Southeast Asia*, Vol. 21, No. 1, April 1999, p. 100.

Swaine, Michael D., "China," in Zalmay Khalilzad (ed.), *Strategic Appraisal 1996*, RAND, MR-543-AF, 1996, pp. 185–221.

Tan, Andrew T.H., "Singapore's Defence: Capabilities, Trends, and Implications," *Contemporary Southeast Asia*, Vol. 21, No. 3, December 1999.

U.S.-ASEAN Business Council, *ASEAN Market Overview*, June 1999.

U.S. Department of Energy, *South China Sea Region*, Washington, D.C.: Energy Information Administration, August 1998.

United States Institute of Peace, *The South China Sea Dispute: Prospects for Preventive Diplomacy*, Washington, D.C., 1996.

Valencia, Mark J., *China and the South China Sea Disputes*, Adelphi Paper 298, Oxford: Oxford University Press for IISS, 1995.

_____, "Energy and Insecurity in Asia," *Survival*, Vol. 39, No. 3, Autumn 1997, pp. 85–106.

Wilhelm, Alfred D., Jr., *China and Security in the Asian Pacific Region Through 2010*, Alexandria, VA: Center for Naval Analyses, March 1996.

Winnefeld, James A., et al., *A New Strategy and Fewer Forces: The Pacific Dimension*, RAND, R-4089/2-USDP, 1992.

Wortzel, Larry M., "China Pursues Traditional Great-Power Status," *Orbis*, Spring 1994, pp. 157–175.

_____, *The ASEAN Regional Forum: Asian Security Without an American Umbrella*, Carlisle Barracks, PA: Strategic Studies Institute, U.S. Army War College, 1995.

Made in the USA
Monee, IL
19 May 2022

96702204R00063